Ecotopia

BGL | David Brooks | Dagmara Genda | Rodney Graham | Isabelle Hayeur
Tristram Lansdowne | Maude Léonard-Contant | Lynne Marsh
Lisa Sanditz | Jennifer Steinkamp | T & T | Kate Wilson

Curated by Amanda Cachia

Kitchener-Waterloo Art Gallery | Southern Alberta Art Gallery | Nickle Galleries | Kenderdine Art Gallery | College Art Galleries

Published by the Kitchener-Waterloo Art Gallery, Kenderdine Art Gallery | College Art Galleries, Nickle Galleries and the Southern Alberta Art Gallery in conjunction with the exhibition *Ecotopia* curated by Amanda Cachia.

Library and Archives Canada Cataloguing in Publication

Cachia, Amanda, 1978-
Ecotopia / by Amanda Cachia, Ernest Callenbach, Anthony Vidler.

Co-published by: Southern Alberta Art Gallery, Nickle Galleries and Kenderdine Art Gallery.

Includes bibliographical references.
Catalogue of an exhibition held at the Kitchener-Waterloo Art Gallery, Kitchener, Ont. from Sept. 28, 2012 - Jan. 6, 2013 and then traveling to other galleries.
ISBN 978-1-897543-20-7

1. Art, Canadian--21st century--Exhibitions. 2. Art, American--21st century--Exhibitions. 3. Environmental protection in art--Exhibitions. 4. Environmental degradation--Exhibitions. I. Callenbach, Ernest II. Vidler, Anthony III. Kitchener-Waterloo Art Gallery IV. Southern Alberta Art Gallery V. Kenderdine Art Gallery VI. Nickle Galleries

VII. Title.

N6545.6.C33 2013 704.9'4933372074 C2013-900707-5

Copy-Editing and Proofreading: Shannon Anderson
Installation Photography on pages 6, 8-17, 29, 31, 33, 44-45, 63: Robert McNair
Graphic Design: Matt Dupuis
Pages 34-35: Rodney Graham, Ponderosa Pines, Princeton B.C./ CAT hi-way yellow, 1992.
Image courtesy of the Vancouver Art Gallery.
Collection of Vancouver Art Gallery, Gift of Sandra Simpson, VAG 98.65.7e
Page 26: Isabelle Hayeur, still from *Losing Ground*, 2009. Image courtesy of the artist and Galerie Division, Montreal.

Printed in Canada by Friesens
Distribution: ABC Art Books Canada
All installation images courtesy of the Kitchener-Waterloo Art Gallery.
COVER IMAGE: Dagmara Genda, *Panorama* (detail), 2012. Photo: Robert McNair.

Published to accompany the exhibition *Ecotopia* at the following venues:

September 28, 2012 – January 6, 2013
Kitchener-Waterloo Art Gallery
101 Queen St. North
Kitchener, Ontario
N2H 6P7

February 9 – April 14, 2013
Southern Alberta Art Gallery
601 Third Avenue South
Lethbridge, Alberta
T1J 0H4

June 14 – July 27, 2013
Nickle Galleries
410 University Court N.W.
Calgary, Alberta
T2N 1N4

January 24 – May 7, 2014
Kenderdine Art Gallery | College Art Galleries
University of Saskatchewan
107 Administration Place
Saskatoon, Saskatchewan
S7N 5A2

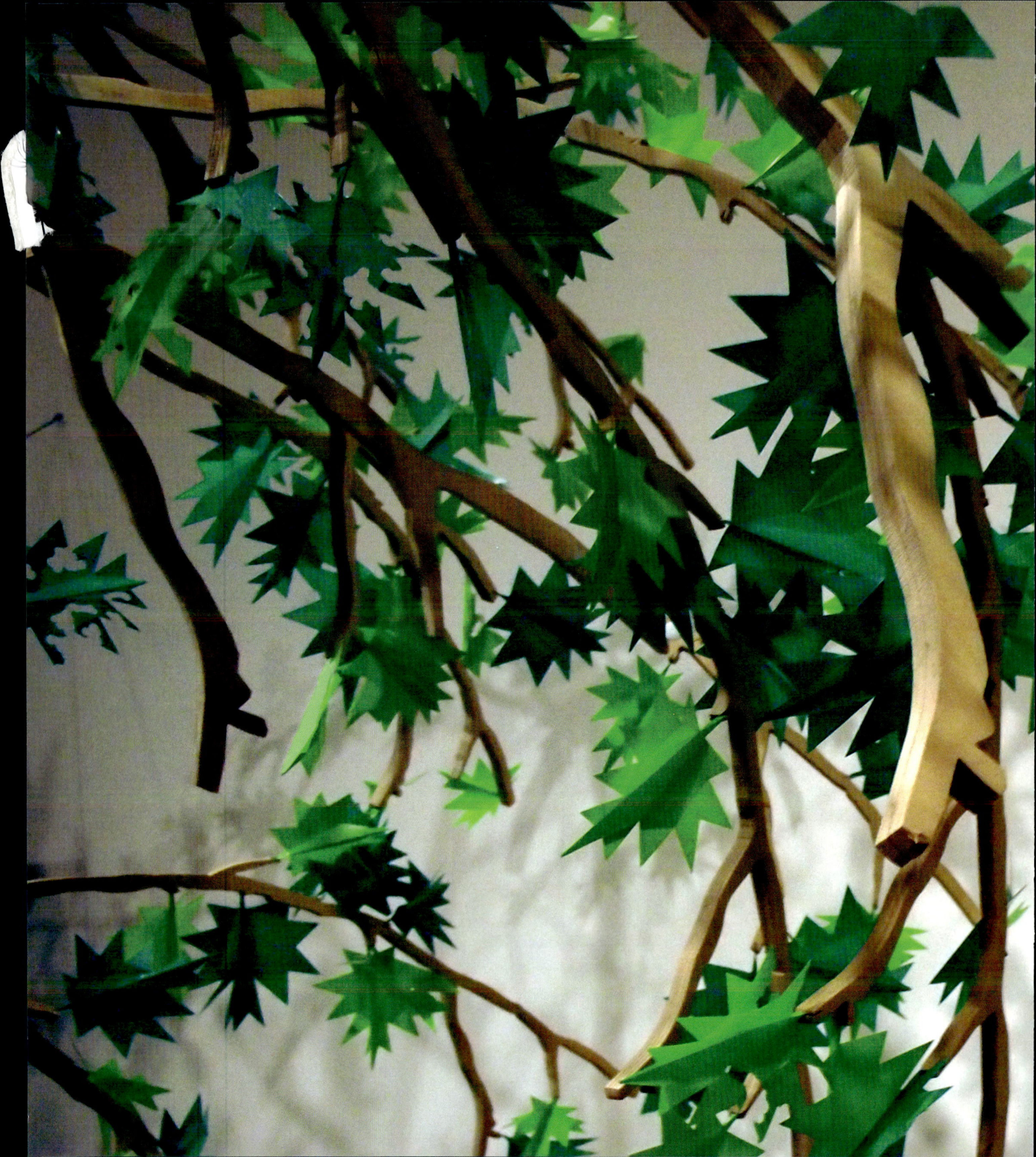

Foreword

Shirley Madill, Executive Director, Kitchener-Waterloo Art Gallery

Often visual art is perceived as separate from the scientific disciplines, yet science and the environment frequently figure prominently in the work of many contemporary artists. Artists have been approaching environmental issues in provocative, diverse and engaging ways for decades, from their choice of media, to the subject matter they address, to the intentions of their work. *Ecotopia* is an exhibition that raises questions about the role of technology in environmental change and its impact on cities and the natural landscape, from the destructive to the transformative. It reveals the role that visual art can play in environmental discourse, particularly at a time of major urban developments. Today, urban areas are hot spots that drive environmental change at rapid speed.

Material demands of production and human consumption alter land use and urban ecology integrates natural and social sciences to study these radically altered environments. In other words, cities present both the problem and solution to an increasingly urbanized world. This exhibition is representative of the Kitchener-Waterloo Art Gallery's programming objectives to create awareness and engage our publics in discussions around contemporary issues through group exhibitions, significant publications and public programming.

I would like to take this opportunity to express our appreciation to the Guest Curator Amanda Cachia on *Ecotopia* and its presentation at KW|AG. I also extend my thanks and congratulations to all the artists who participated and opened up critical questions concerning our urban and environmental future. It is an honour to have their works presented at KW|AG. I also would like to express my deepest appreciation to all the lenders to the exhibition.

Very special commendation goes to KW|AG's Senior Curator Crystal Mowry for her hard work and aesthetic vision in facilitating this presentation and to all KW|AG staff whose input at every level ensures professionalism and quality in all that we do.

With its emphasis on the shifting dynamic between urban and rural experiences, *Ecotopia* is an exhibition that speaks to many Canadians. We are pleased to have partnered with the Southern Alberta Art Gallery in Lethbridge, the Nickle Galleries in Calgary, and the Kenderdine Art Gallery | College Art Galleries in Saskatoon on the publication and tour of this exhibition.

Ecotopia was sponsored by The Musagetes Fund at The Kitchener and Waterloo Community Foundation and I express our gratitude for this support.

On behalf of our publishing partners we express our deepest appreciation to our major funders: the Canada Council for the Arts, the Ontario Arts Council, the Alberta Foundation for the Arts, the Saskatchewan Arts Board, SaskCulture, the City of Lethbridge, the University of Saskatchewan, the City of Kitchener and the City of Waterloo.

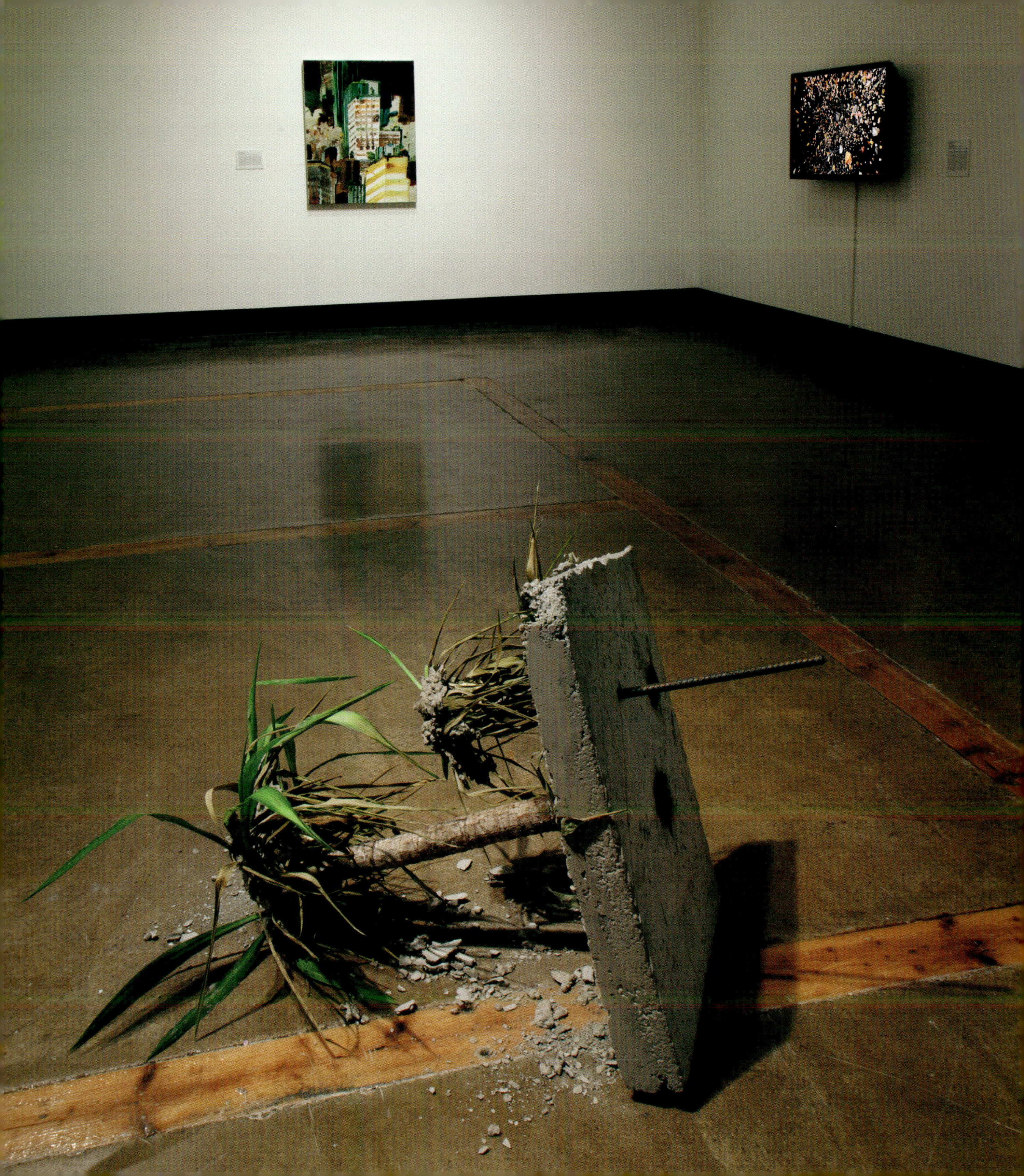

Second Nature

Amanda Cachia, Exhibition Curator

Cicero called nature "second nature" or nature transformed by the human hand.[1]

The country and the city are traditional opposites: the rural landscape is spacious, flowing, totalizing, and spiritually uplifting; the cityscape is crowded, sharp-edged, fragmented and morally corrupting. That tradition is being challenged. The city, more and more, is understood as an ecosystem, and the country is allowed corruptions and constructions of its own. Neither could exist without the other, and understanding their deep connectedness makes it harder to uphold the polarity.[2]

We didn't just invent the ideal of beauty; much of what we consider beautiful we shaped ourselves.[3]

Ecotopia explores environmental conservation, destruction and the discordant combination of architecture and decay in our technological age.[4] Very simply, this exhibition is about the confluence of nature and human intervention that has both glorious and profoundly dispiriting qualities. These seemingly contradictory elements are captured in the title of this exhibition, for the word "ecotopia" can be either utopian or dystopian, where a vision for a world is either ideal or nightmarish.[5] These jarring notions are brought into conversation in the work of twelve artists who offer alternatives for postmodern living that enable a harmonious coexistence of nature and technology rather than environmental havoc. The artists also ponder the expressionistic grandeur of nature, and how this has become suffocated and overladen with layers of city debris and decaying architecture. Visitors will come across sculptures, paintings, works on paper and video installations that address various aspects of history, technology, geography, Canadian national identity, politics, urbanization and ecology. The artists approach the natural and built environment with anxiety, humour, irony, amazement and terror.

Architectural and cultural theorist and historian Anthony Vidler has developed numerous relevant discourses around the origins and affective consequences of modern spaces:

Fear, anxiety, estrangement, and their psychological counterparts, anxiety neuroses and phobias, have been intimately linked to the aesthetics of space throughout the modern period. Romanticism, with its delight in the terrifying sublime, saw fear and horror lurking in landscapes, domestic scenes, and city streets. Modernism, while displacing many such spatial fears to the domain of psychoanalysis, was nevertheless equally subject to fears newly identified as endemic to the metropolis...calculating its modes of representation according to the psychological disturbances of an alienated subject. Space, in these various iterations, has been increasingly defined as a product of subjective projection and introjection, as opposed to a stable container of objects and bodies.[6]

This "second nature" is suspect; it is manipulative, beautiful and horrific all at once according to the projections of the

1. Ginger Strand, "At the Limits: Landschaft, Landscape and the Land," in *Badlands: New Horizons in Landscape*, ed. Denise Markonish (North Adams, MA: MASS MoCA, 2008), 82.
2. Ibid., 86.
3. Ibid., 82.
4. "Eco" is from the Greek word *oikos* (household or home) and "topia" is from the Greek word *topos* (place). Ernest Callenbach, *Ecotopia: The Notebooks and Reports of William Weston* (Berkeley, CA: Banyan Tree Books, 1975).
5. The term was originally developed by ethnographer E.L. Anderson and popularized by the late American writer Ernest Callenbach in his novel of the same name, where ecotopia became a subgenre of science fiction. The society described in the book is one of the first ecological utopias and was influential on the counterculture, and the green movement in the 1970s and thereafter. It was a protest against consumerism and materialism, among other aspects of American life.
6. Anthony Vidler, "Introduction" in *Warped Space: Art, Architecture, and Anxiety in Modern Culture* (Cambridge, MA: MIT Press, 2000), 1.

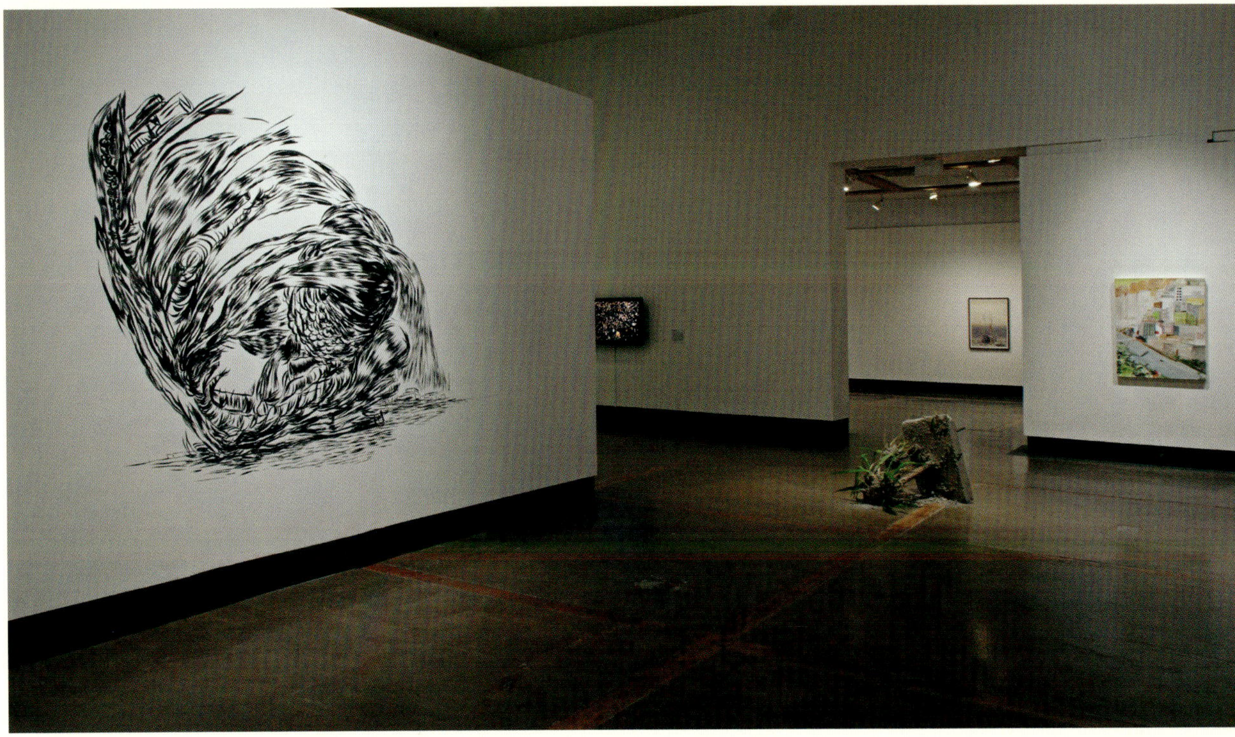

7. United Nations website, http://www.un.org/en/events/iyof2011/ (accessed December 3, 2012).

artists in this exhibition regarding the inhabitance of our contemporary space. Any boundaries crossed by the artists create discomfort that moves and stirs the viewer to understand that what is at stake is not simply what we think of as "beauty" but our way of moving forward in our flawed world.

The genesis for this exhibition was the 2011 International Year of Forests, declared by the United Nations as "a global celebration of people's action for sustainable forest management."[7] For many people, the icon of the tree symbolizes nature and the environment in its totality. It also represents the rehabilitated landscape for which we yearn. Indeed, the environment and our landscape is a construct of the mind, memory and political will as much as the physical configuration of natural and built spaces.

The tree has also been of great inspiration for contemporary artists around the world. In 1982, the German Fluxus and performance artist Joseph Beuys began a large-scale seminal project in Kassel, Germany as part of *documenta 7* entitled *7000 Oaks*. His plan was to plant 7000 saplings throughout the city of Kassel, and it took over five years to complete, concluding with the opening of *documenta 8* in 1987. Beuys meant for his project to be just the start of an ongoing tree-planting effort extended throughout the world as part of a global mission to affect environmental and social change on a mass scale and to raise ecological consciousness. Within the city of Kassel itself, Beuys wanted to make a gesture toward urban renewal. German artist and author Johannes Stüttgen said that "the planting of seven thousand oaks is thus only a symbolic beginning...

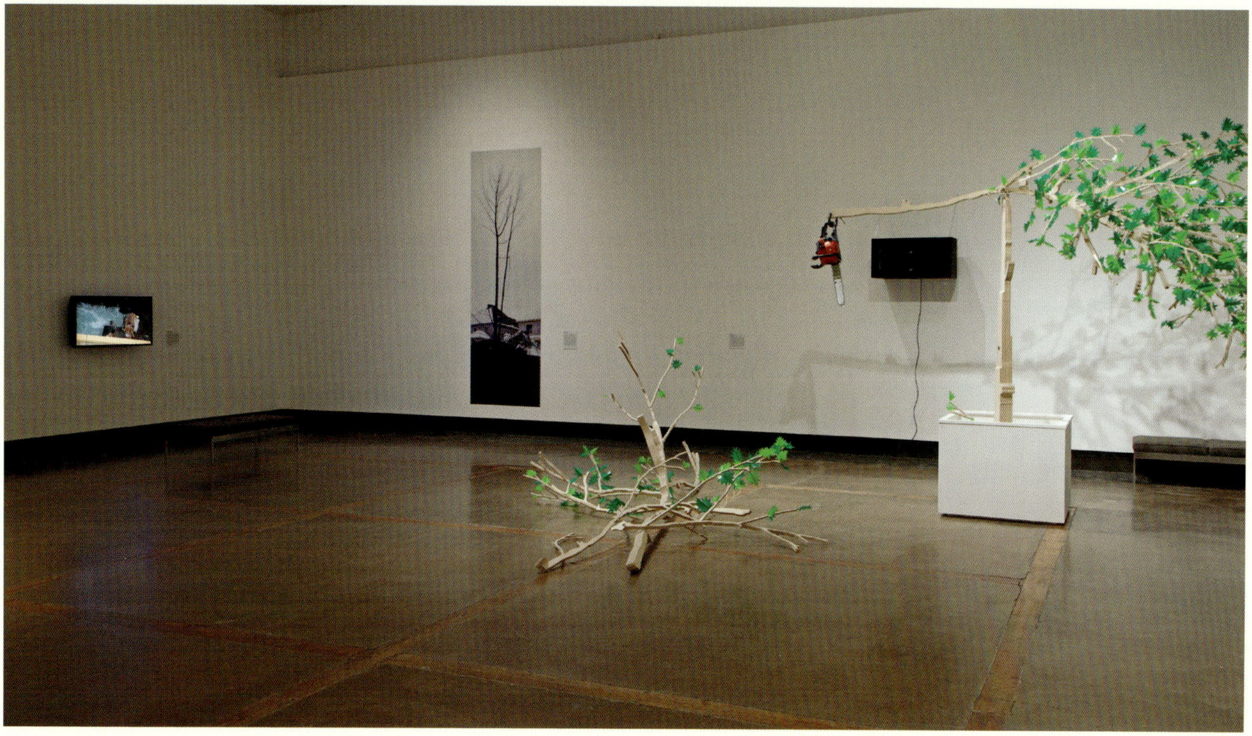

The intention of such a tree-planting event is to point up the transformation of all of life, of society, and of the whole ecological system."[8] Beuys felt an urgency to act out his anxiety regarding environmental issues by merging visual art, social practice and environmental activism. *7000 Oaks* was his utopian vision of a social sculpture that was designed to provoke revolution, permanence and longevity.

Interspersed throughout *Ecotopia* are multiple and diverse representations and readings on trees. Each one suggests the adaptive potential and resiliance of trees in the wake of this "second nature" we live in. For example, BGL's artificial tree is made up of machine-cut plywood with plastic leaves (some hole-punched as though eaten by worms) spinning in a circle caused by the wind from several electric fans (*Pinocchio*, 2009); David Brooks' sculpture includes two small palm trees that seem to have emerged from the sidewalk, now caked in concrete and lying desolate on their side (*Unfinished Section of Sidewalk with Two Palms and Rebar*, 2012); Tristram Lansdowne's watercolour paintings of trees are life and death personified—lush, velvet, green leaves devouring a building in one image; spindly, dried up branches in a desolate grey landscape in the other (*Envelope*, 2012 and *The Nurse*, 2011); finally, Jennifer Steinkamp's computer-generated trees and leaves constantly orbit across a black space and mutate from shape to shape, colour to colour, evoking seasons, and more profoundly, change (*Orbit 6*, 2010). What has life become if we are now mesmerized by the movement of plastic trees or those we watch on a television screen instead of a genuine breeze flowing through real trees in our parks and forests?

8. Johannes Stüttgen, *Beschreibung eines Kunstwerkes* (Düsseldorf: Free International University, 1982), 1.

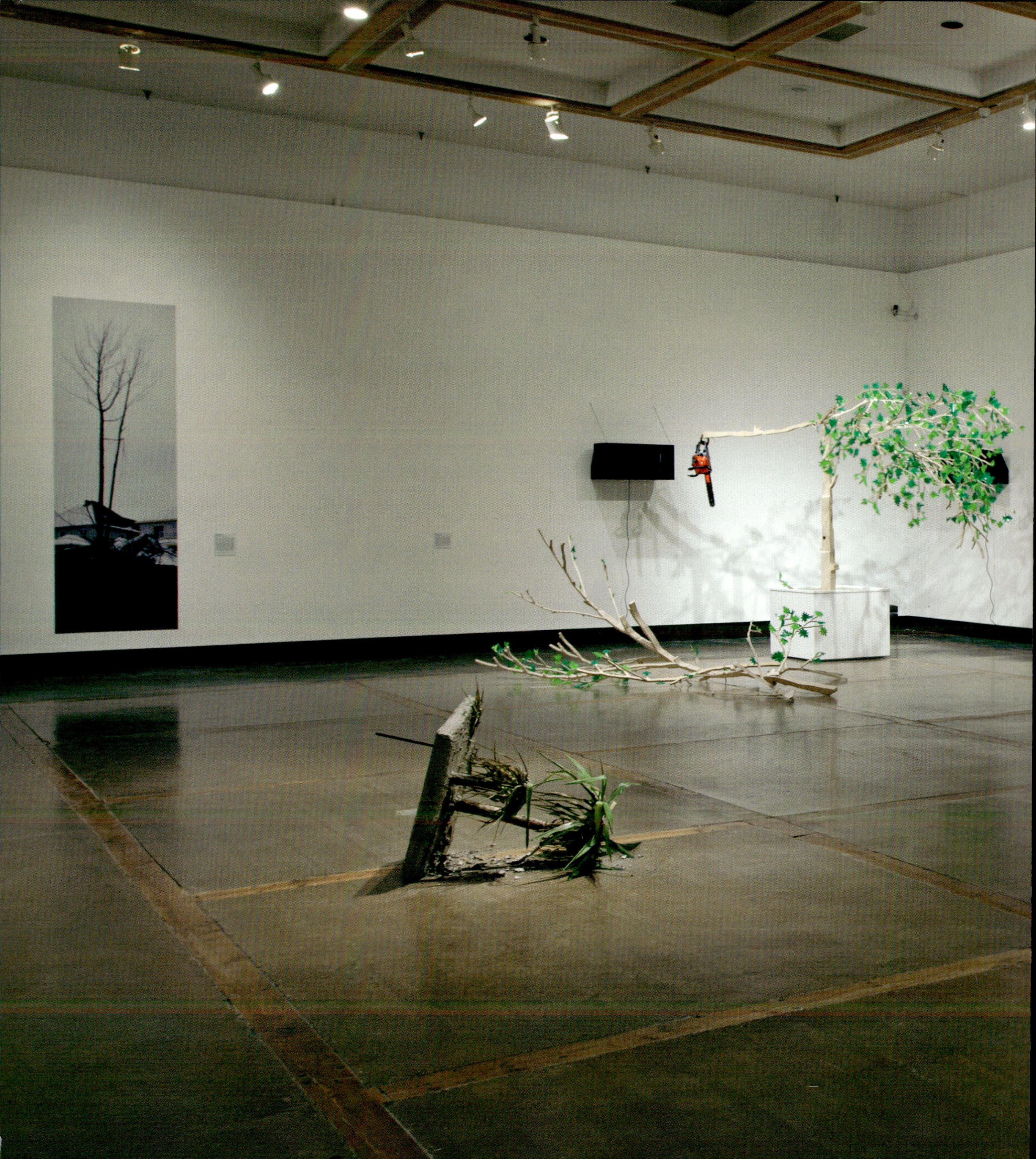

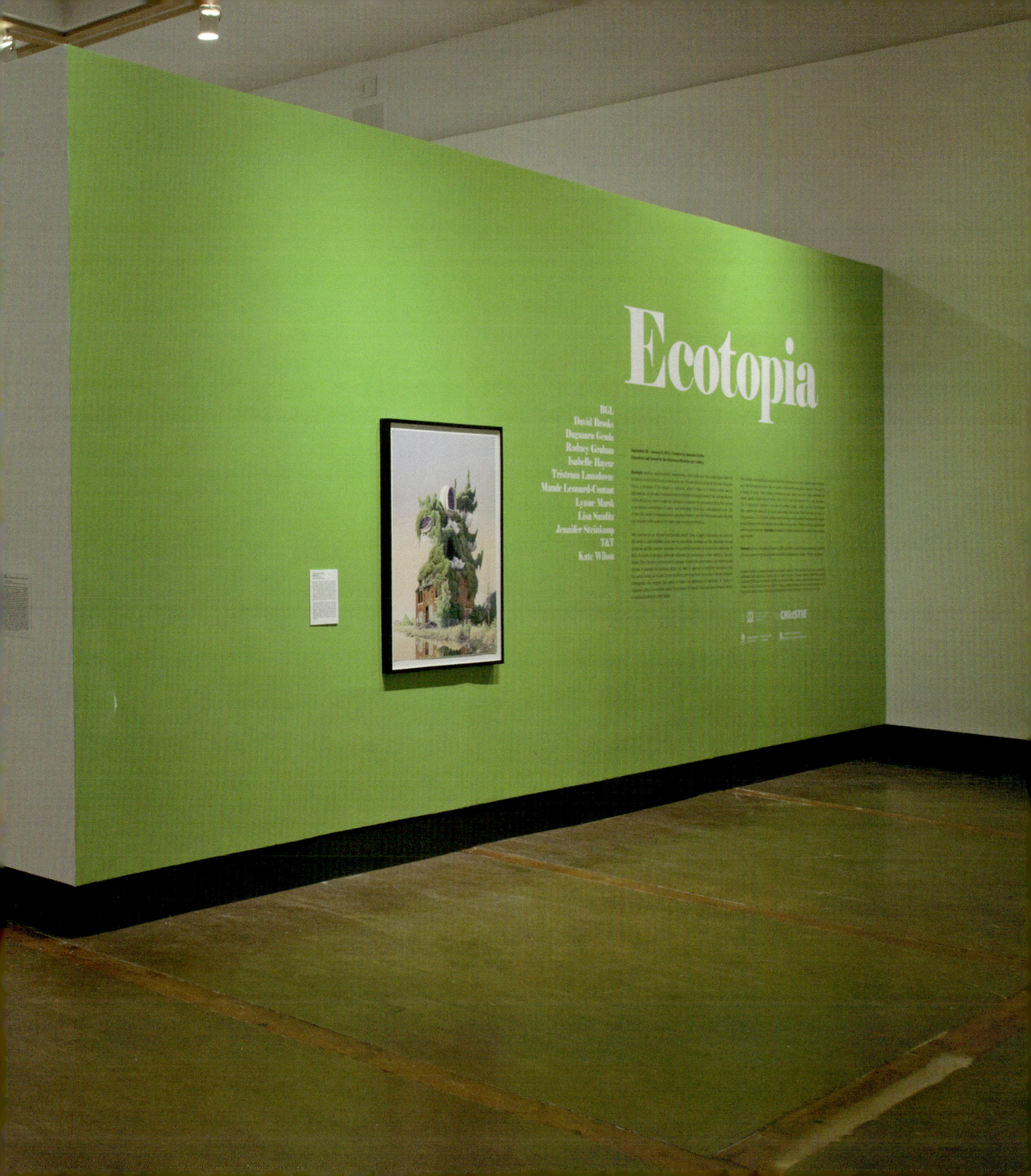
Ecotopia
BGL
David Brooks
Dagmara Genda
Rodney Graham
Isabelle Hayeur
Tristram Lansdowne
Maude Léonard-Contant
Lynne Marsh
Lisa Sandlitz
Jennifer Steinkamp
T&T
Kate Wilson

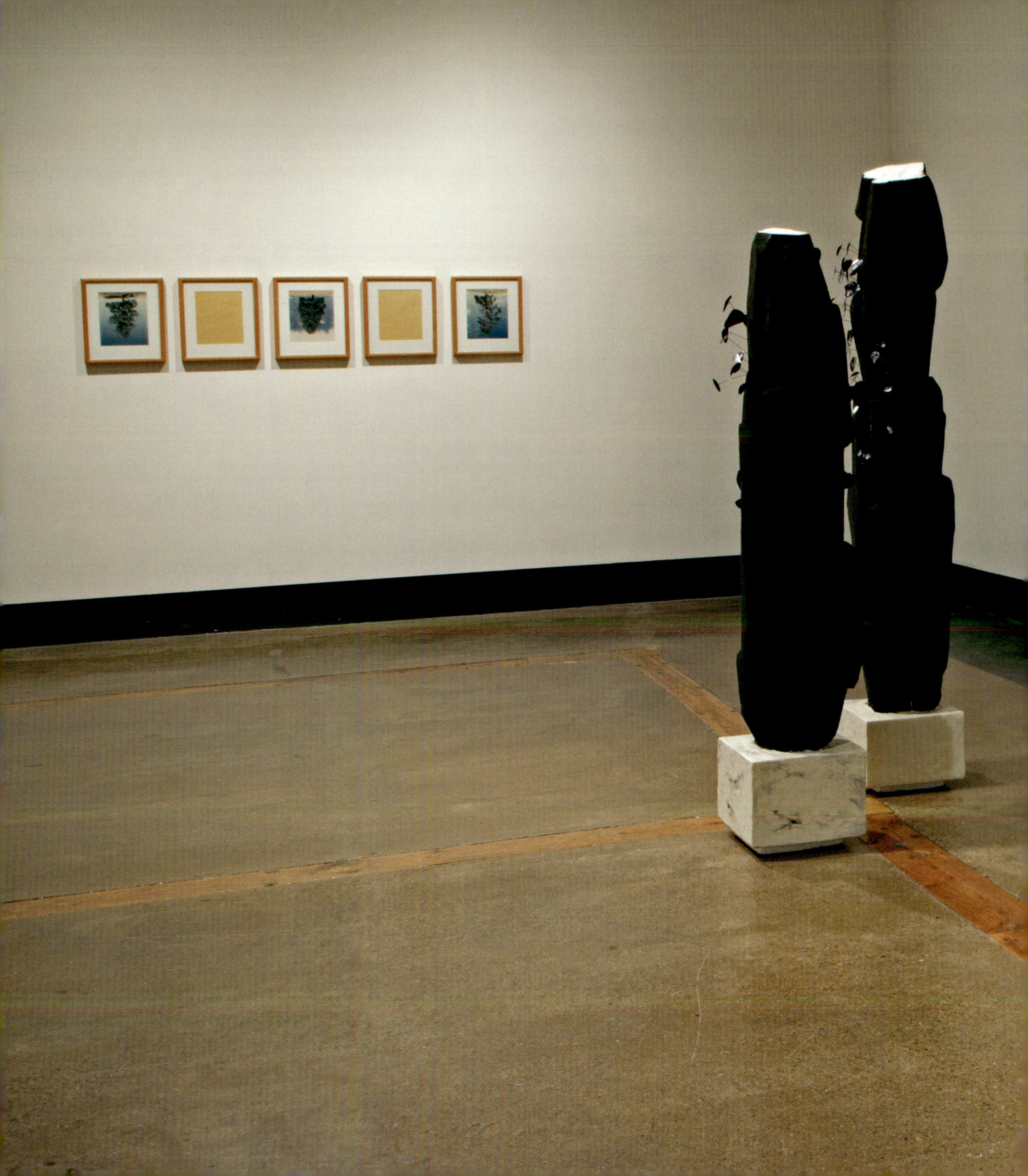

Rodney Graham suggests that perhaps the world—and perceived romantic notions of nature, also touched on by some of the artists—has been turned on its head, as evoked in his upside-down Ponderosa Pines (*Ponderosa Pines, Princeton B.C./ CAT hi-way yellow*, 1992).

If Cicero called nature "second nature" or nature transformed by the human hand, then why and how do we recreate and mediate nature? Predominantly, we live in abstract spaces and manufactured environments where nature has become fabricated—consider the tightly manicured lawns and rows of young trees that mark most cookie-cutter housing developments, for instance. Using a variety of strategies and materials, some of the artists in this exhibition point out the absurdities, excesses and challenges to these situations, and the amnesic response our population has taken up to these new falsehoods—these ecotopias. Other artists suggest that while our world may be topsy-turvy, there might be a beauty of decay. For example, the new and decaying structures that we have created have a suggested beauty in themselves, where objects, monuments, and sites that have been overtaken by weeds, graffiti, rabbit holes or worms are in fact a new archaeology for future generations. With its ability to perplex and entice in equal measures, *Ecotopia* unsettles comfortable notions about dominion and progress.

Our understanding of the term "landscape" has become more complex over time. The tradition of landscape painting, for instance, has evolved dramatically since the days of bucolic, colonial scenes of sheep grazing in meadows amidst the grandeur of nature. American writer and novelist Ginger Strand captures this evolution in the following statement:

No one believes in landscape anymore. As a self-contained genre, pretty vistas and sublime scenes feel compromised. There's a shadow moving across those sylvan fields, the shadow of ideology...Landscape masqueraded for several centuries as objective, a window onto the natural world. That's all over now. Today, like so much else, landscape has been unmasked.[9]

The work of Dagmara Genda and Maude Léonard-Contant touches on the essence of this quote, for their works allude to how the pretty vistas in Canadian national parks are actually constructed through particular ideological frames or windows of a now compromised landscape. For example, Dagmara Genda's *Panorama* (2012) is a 360° paper mural of Canada and is made from collaged images sourced from nature books and travel brochures. The artist says, "The work refers to signs and regulations in national parks and spots where people come to enjoy the wilderness of Canada and I am interested in the 'ownership' of wilderness and how that has become a cultural commodity."[10] In Maude Léonard-Contant's *Creeper* (2012), *Cacti* (2012) and *Coco-fesse* (2011), the artist says that she is "exploring strategies of simulation and camouflage, blurring the boundaries between fake and real, between interior design and landscape architecture."[11] What kind of aesthetics come into play, then, in the constructed environments revealed in the work Genda and Léonard-Contant? Perhaps it is "ruins aesthetics"? UK-based writer and critic Brian Dillon says that in recent years, many artists have turned to themes of destruction and decay in their works, with a particular focus on the remains of buildings and landscapes that seem like relics of the past century. He says that "the ruin appears in its material specificity as an allegory for global or regional political forces, as an aesthetic trope that condenses images and ideas from different genres, as an invitation to thought and as the site of a sublime...haunting."[12]

9. Strand, "At the Limits: Landschaft, Landscape and the Land," 81.
10. Dagmara Genda artist statement, 2012
11. Maude Léonard-Contant artist statement, 2012
12. Brian Dillon, "Introduction: A Short History of Decay," in *Ruins (Whitechapel: Documents of Contemporary Art)*, ed. Brian Dillon (Cambridge, MA: MIT Press, 2011), 11.

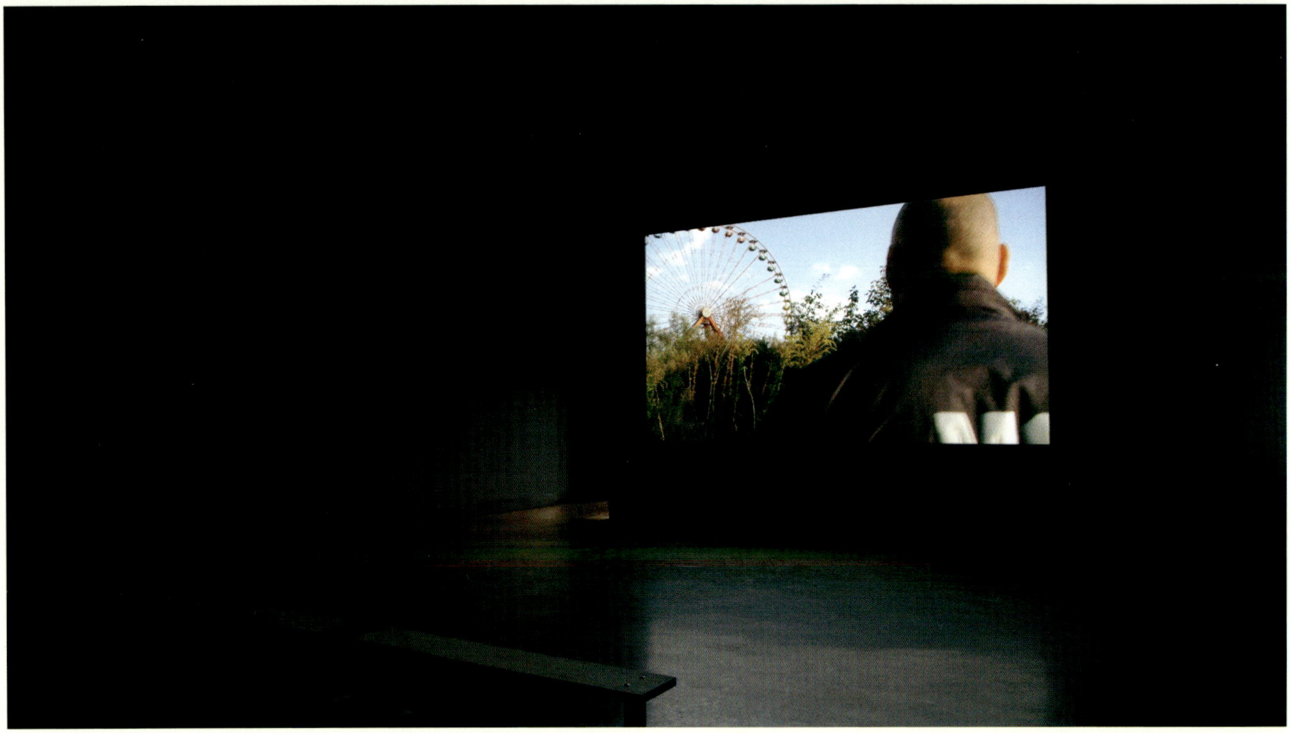

13. Ibid., 14.
14. Magali Arriola, "A Victim and a Viewer: Some Thoughts on Anticipated Ruins," in *Ruins (Whitechapel: Documents of Contemporary Art)*, 176.

Like the paradoxical term "ecotopia," the ruin also embodies a set of temporal and historical paradoxes. The ruined building is a remnant and portal into the past, and its decay is a reminder of the passage of time. And yet, despite being a ruin, it somehow survives and outlives us, representing who and what we are at any given moment for future generations to uncover and decipher. To adopt a term from the Land Art artist Robert Smithson, "ruins in reverse" is where ruins are caught in a dialectical state between being built and falling into disuse and decay.[13] Within this peculiar state, ruins slowly accumulate signs of transformation in the urban landscape over time, as if they are markers or time capsules for events. For example, tourism transforms and further degrades famous sites containing important ruins, such as Machu Picchu in Peru or the pyramids in Egypt. There has even been talk of re-building some of the architecture at the Auschwitz concentration camp in Poland. What are the ethical and social implications of re-building a ruinous site?

Visiting and revisiting ruins and sites of decay gives viewers the opportunity to place new significance on the stratification of remnants and recover narratives that inhabit these spaces of the past in order to engage their present life and function.[14] In the context of the exhibition, Lynne Marsh's *Plänterwald* (2010) and Kate Wilson's *Untitled* (2012) large-scale ink drawing reference places that have become undone over time, either through natural forces or man-made constructions. Marsh has filmed on the site of a former German Democratic Republic amusement park built in 1969 and abandoned after unification. Its rollercoaster and Ferris wheel sit motionless at the edge of the city of Berlin. Wilson has depicted a frightening

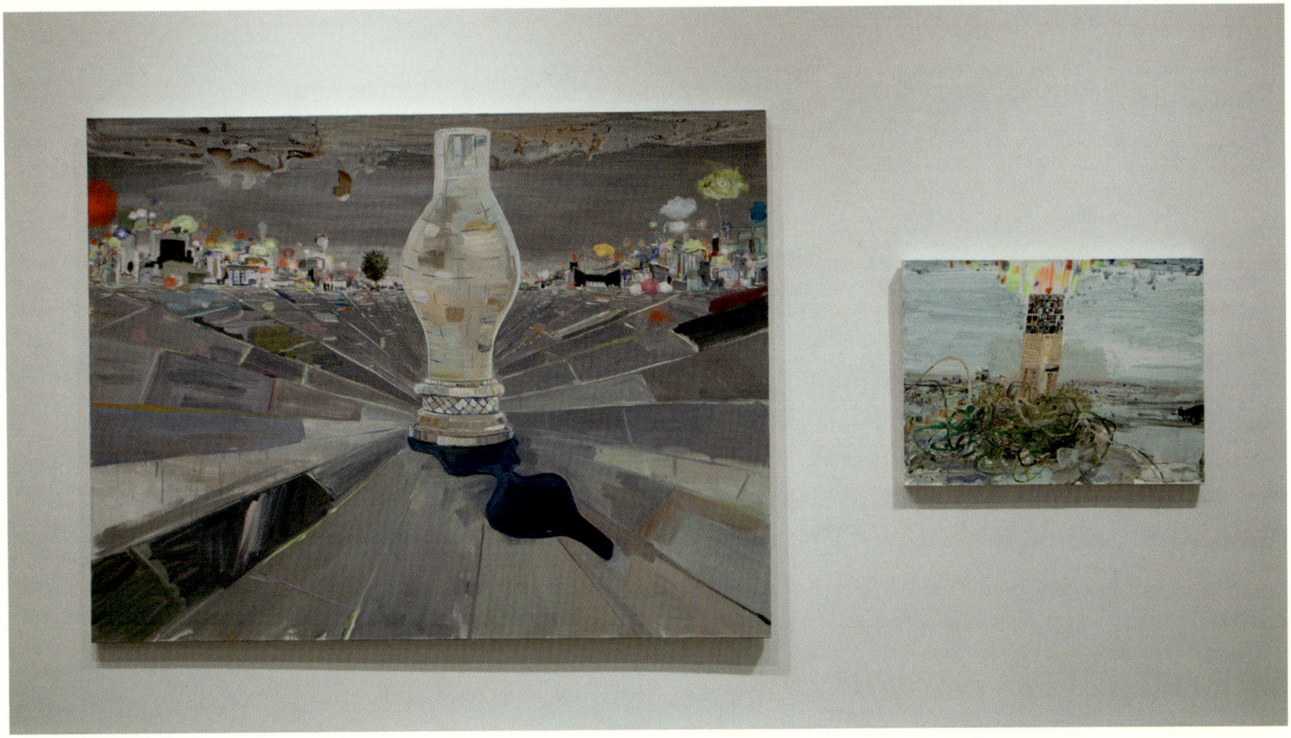

scene where a botanical swirl is caught up in a hurricane-like wind, sucking up nearby architecture and other structures and elements in nature that line the horizon. What do the markings of these spaces reveal about those who currently frequent or even live in them?

The work by the artists in this exhibition can be considered a type of resistance against the regulation and homogenization of public architecture and spaces in our environment. Artists such as Isabelle Hayeur (*Ascendance*, 2005-2008, and *Losing Ground*, 2009), Lisa Sanditz (*Lamp City*, 2009, *Sock City*, 2005, *Tang Factory I*, 2008 and *Tower of Babble*, 2009) and T&T (*Wilmont Trail*, 2003) examine how local suburban and rural realities are increasingly marked by identical housing developments, retail complexes, shopping arcades and urban villages. In doing so, they aim to reveal how the dominant forms of urban planning and development are often in conflict and incongruence with nature. Curator Magali Arriola says that such spaces become "conflicted and contested arenas that simultaneously contain traces of their aspirations as they carry the marks of their failure."[15]

Furthermore, the artists are creating and inhabiting new spaces of geographic "otherness" that French philosopher Michel Foucault termed "heterotopias."[16] According to Foucault, a heterotopic space could be simultaneously physical and mental, such as the moment you see yourself in the mirror or the space of a phone conversation. While many of the post-apocalyptic "places without places" evident in the artists' works are fueled by the imagination, each artist has also been effected by the sights they behold in their daily routines

15. Ibid.

16. Michel Foucault, "Of Other Spaces," in *Power/Knowledge: Selected Interviews and Other Writings, 1972–1977*, ed. Colin Gordon (New York: Vintage, 1980).

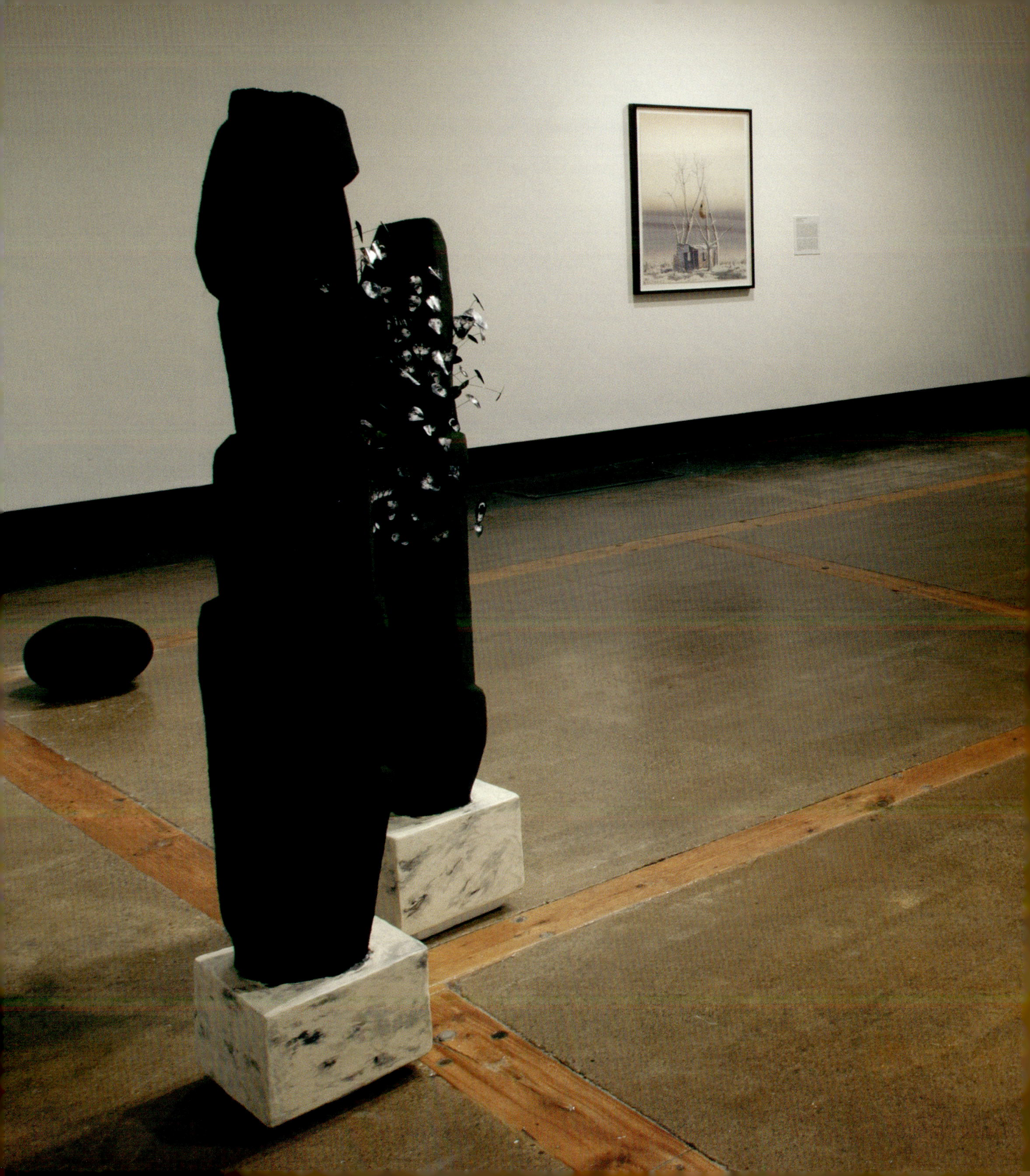

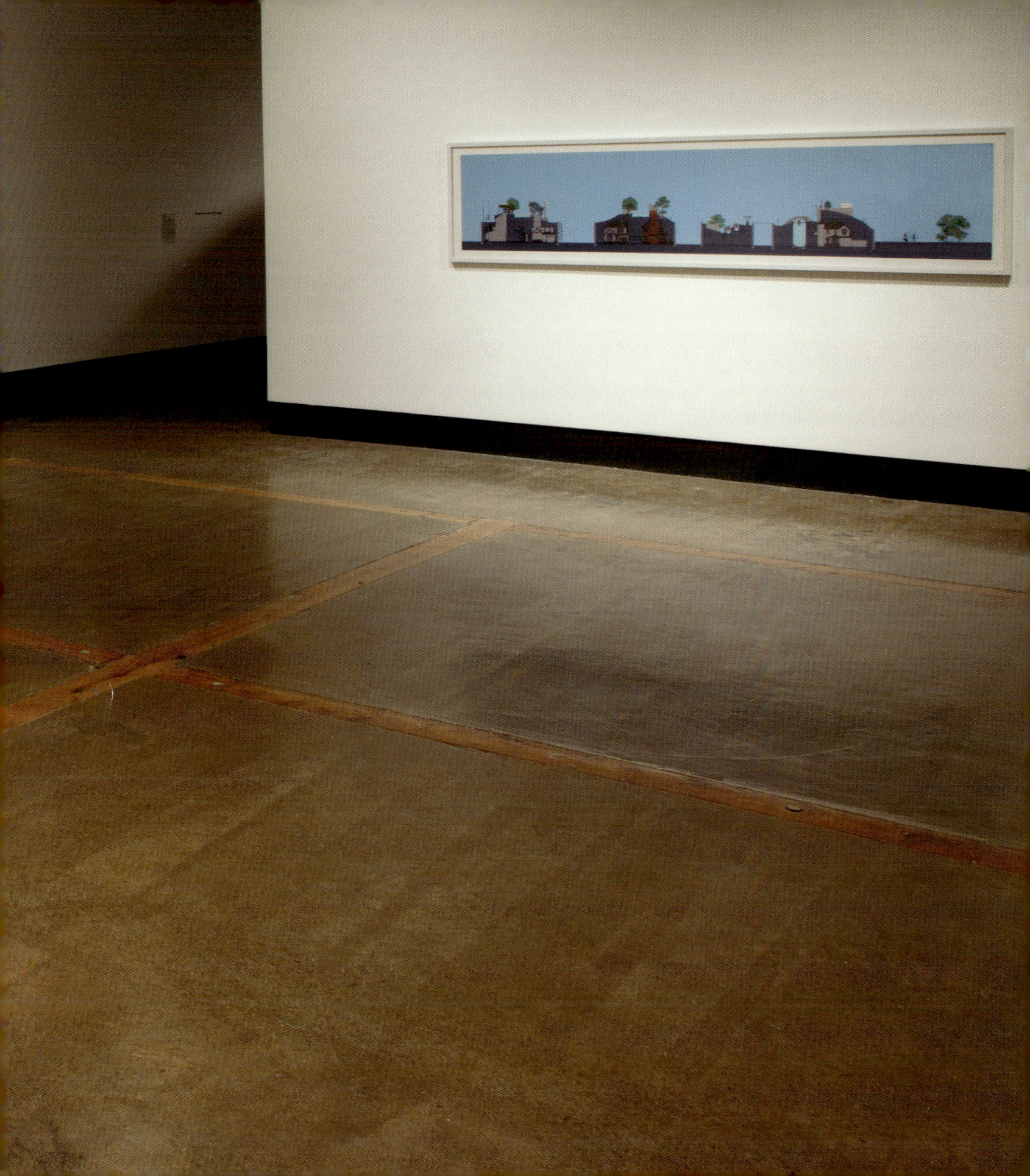

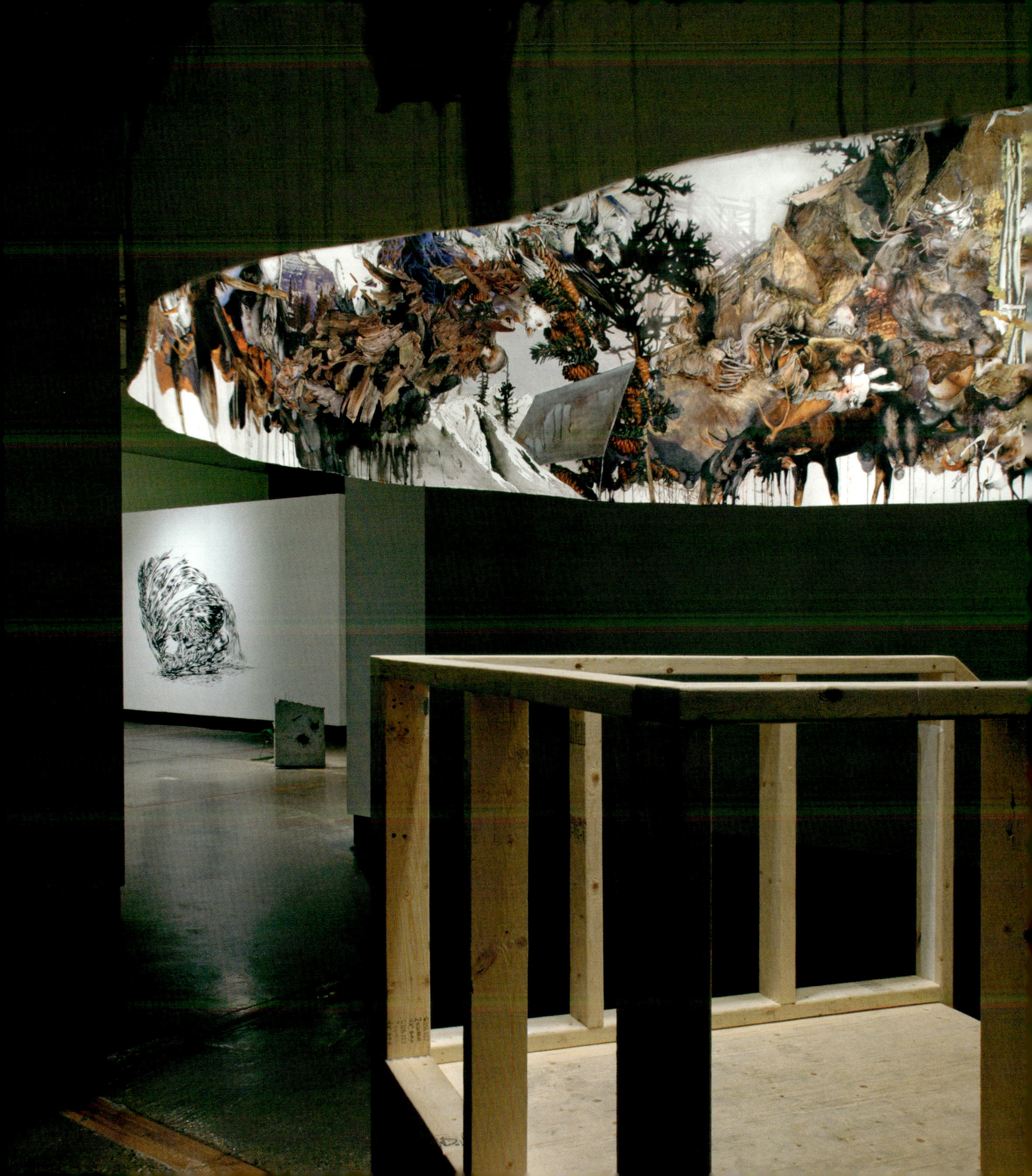

and travels or images they come across in the media or the internet. Within such heterotopic spaces, there is room for escapism, fantasy and play and maybe even hope. Art critic and curator Patrik Andersson said of T&T (which also applies to the other artists in *Ecotopia*) that they "refashion their own potential future in a way that addresses our environmental crisis without letting go of cultural habits born out of urban and suburban modernization."[17] In other words, despite creating places of otherness, ruins in reverse or spaces in between, the artists are still grounded in inescapable everyday living. The twelve artists featured in *Ecotopia* capture the spirit of a world that is changing definitions on multiple levels: terms such as "landscape," "ecosystem" and "nature" evolve as climate change impacts our environment. Nature and city meet to form a vocabulary for modern-day living that makes the world an equally exciting and daunting place to live. What will the world be like for future generations? Will we find ourselves in a world like the one depicted in Dr. Seuss' children's book and film of the same name, *The Lorax*?[18] Will there still be seasons and forests, or even trees? What will be the next big natural disaster and how will the world respond? Will the ruins of the future bear a resemblance to the great ruins of our past, such as Stonehenge in England, or the Colosseum in Rome? Is it possible that failed, abandoned architecture could be rebuilt to create ecologically sustainable spaces? Brian Dillon states that "the ruin is not of melancholy or mourning but of radical potential—its fragmentary, unfinished nature is an invitation to fulfill the as yet unexplored temporality that it contains."[19] It is thanks to the artists and their own ecotopian visions of our world that we can begin to imagine some of the possibilities that may thrill, threaten and shake us.

Amanda Cachia is an independent curator from Sydney, Australia and is currently completing her PhD in Art History, Theory & Criticism at the University of California, San Diego. Her dissertation will focus on the intersection of disability and contemporary art. Cachia completed her second Masters degree in Visual & Critical Studies at the California College of the Arts (CCA) in San Francisco in spring, 2012. Her MA thesis, entitled *What Can a Body Do? Inscribing and Adjusting Experiences of Disability in Contemporary Art* formed the basis of an exhibition curated by Cachia and hosted by Cantor Fitzgerald Gallery at Haverford College, PA in fall, 2012. Cachia received her first Masters in Creative Curating from Goldsmiths College, University of London in 2001. She held the position of Director/Curator of the Dunlop Art Gallery in Regina, Saskatchewan, Canada from 2007-2010, and has curated approximately 30 exhibitions over the last ten years in London, New York, Oakland and various cities across Australia and Canada. Her writing has been published in numerous exhibition catalogues, *Canadian Art* magazine and upcoming issues of *Disability Studies Quarterly* and she has lectured and participated in panels and conferences widely, including the National Gallery of Canada, Winnipeg Art Gallery, University of California (Berkeley, Santa Barbara and San Diego), Paul K. Longmore Institute at San Francisco State University, Portland State University, de Young Museum, Metropolitan Museum of Art and the Graduate Center at City University of New York. Cachia has been the Chair of the Dwarf Artists Coalition for the Little People of America since 2007.

17. Patrik Andersson, "Doubt/Hope," in *T&T: Onward Future* (London: Museum London, Oakville, ON: Oakville Galleries with Trapp Editions, 2008), 12.

18. *The Lorax* is a children's book written by Dr. Seuss and first published in 1971. The environment is in jeopardy in this story as there are no trees left in the world thanks to the greed of humanity, particularly the central character, Once-ler. The Lorax fights for the trees, the environment and against Once-ler. The film of the same name was released by Universal Studios and Illumination Entertainment in 2012.

19. Dillon, "Introduction: A Short History of Decay," 18.

Excerpt from

Ecotopia

Ernest Callenbach

Ecotopia: Challenge or Illusion?

San Francisco, June 19. Where is Ecotopia going in the future? After more than six weeks' intensive study of the country, I find it still hazardous to guess. There is no doubt, I have been forced to conclude, that the risky social experiments undertaken here have worked on a biological level. Ecotopian air and water are everywhere crystal clear. The land is well cared for and productive. Food is plentiful, wholesome, and recognizable. All life systems are operating on a stable-state basis, and can go on doing so indefinitely. The health and general well-being of the people are undeniable. While the extreme decentralization and emotional openness of the society seem alien to an American at first, they too have much to be said in their favor. In these respects, I believe, Ecotopia offers us a difficult challenge, and we have far to go to even approach their achievements.

On the other hand, these benefits have been bought at a heavy cost. Not only is the Ecotopian industrial capacity and standard of consumption markedly below ours, to a degree that would never be tolerated by Americans generally, but the Ecotopian political system rests on assumptions that I can only conclude are dangerous in the extreme. In my earlier columns I described the city-states that have already, in effect, themselves seceded within Ecotopia. There is talk currently of formalizing the Spanish-speaking and Japanese communities of San Francisco—the latter, of course, an economically sinister development because of the threat of Japanese capital taking over. Jewish, American Indian, and other minorities all contain militants who desire a greater autonomy for their peoples.

It is, admittedly, difficult for an American to criticize such trends when our own society, after the failure of the integrationist campaign of the sixties, has grown ever more segregated—though somewhat less unequal. However, it is still the American ideal that all men and women should obtain equal protection from the law and have equal status as citizens of one great and powerful nation. The Ecotopian principle of secession denies this hope and this faith. While seemingly idealistic, it is in fact profoundly pessimistic. And the consequences seem clear. The way propounded by Ecotopian ideologies leads away from the former greatness of America, unified in spirit "from sea to shining sea," toward a balkanized continent—a welter of small, second-class nations, each with its own petty cultural differentiations. Instead of continuing the long march toward one world of peace and freedom, to which America has dedicated itself on the battlefields of Korea, Vietnam, and Brazil (not to mention our own Civil War), the Ecotopians propose only separatism, quietism, a reversion toward the two-bit principalities of medieval Europe, or perhaps even the tribalism of the jungle.

Under Ecotopian ideas, the era of the great nation-states, with their promise of one ultimate world-state, would fade away. Despite our achievements of a worldwide communications network and jet travel, mankind would fly apart into small, culturally homogenous groupings. In the words of Yeats (an early 20th-century poet from Ireland—a very small and secessionist country): "The center cannot hold."

Ecotopians argue that such separatism is desirable on ecological as well as cultural grounds—that a small regional society can exploit its "niche" in the world biosystem more subtly and richly and efficiently (and of course less destructively) than have the superpowers. This seems to me, however, a dubiously fetishistic decentralism. It assumes that the immensely concentrated resources of the superpowers are innately impossible to use wisely. I would be the last to deny that the huge administrative machines of our governments and international corporations do not commit an occasional error, or miss an an occasional opportunity. Nevertheless, to condemn them and eliminate them, in favor of small-scale innovations modeled on the Ecotopian experience, would seem to risk throwing the baby of civilization itself out with the polluted bathwater. If we wish to achieve better living conditions for ourselves and our descendents, surely the wiser utilization of the methods we know best is the only way to accomplish it.

Ernest Callenbach (April 3, 1929 – April 16, 2012) was an American writer and editor best known for his visionary novel *Ecotopia* (1975), a best-selling environmental classic. Born and raised in rural central Pennsylvania, he attended the University of Chicago. He migrated to California and joined the staff of the University of California Press. There he founded and edited the internationally known journal Film Quarterly. Concurrently, he edited the UC Press's world-renowned list of film books as well as books in art and science, including the California Natural History Guides series.

Excerpt from the Introduction to

The Architectural Uncanny: Essays in the Modern Unhomely

Anthony Vidler

1. Walter Benjamin, Charles Baudelaire: *A Lyric Poet in the Era of High Capitalism*, translated from the German by Harry Zohn (London: New Left Books, 1973), 128-131: "Fear, revulsion, and horror were the emotions which the big city crowd aroused in those who first observed it" (131).

The contemporary sensibility that sees the uncanny erupt in empty parking lots around abandoned or run-down shopping malls, in the screened trompe l'oeil of simulated space, in, that is, the wasted margins and surface appearances of postindustrial culture, this sensibility has its roots and draws its commonplaces from a long but essentially modern tradition. Its apparently benign and utterly ordinary loci, its domestic and slightly tawdry settings, its ready exploitation as the *frisson* of an already jaded public, all mark it out clearly as the heir to a feeling of unease first identified in the late eighteenth century.

Aesthetically an outgrowth of the Burkean sublime, a domesticated version of absolute terror, to be experienced in the comfort of the home and relegated to the minor genre of the *Märchen* or fairy tale, the uncanny found its first home in the short stories of E.T.A. Hoffmann and Edgar Allan Poe. Its favorite motif was precisely the contrast between a secure and homely interior and the fearful invasion of an alien presence; on a psychological level, its play was one of doubling, where the other is, strangely enough, experienced as a replica of the self, all the more fearsome because apparently the same.

At the heart of the anxiety provoked by such alien presences was a fundamental insecurity: that of a newly established class, not quite at home in its own home. The uncanny, in this sense, might be characterized as the quintessential bourgeois kind of fear: one carefully bounded by the limits of real material security and the pleasure principle afforded by a terror that was, artistically at least, kept well under control. The uncanny was, in this first incarnation, a sensation best experienced in the privacy of the interior...

But the uncanny, as Walter Benjamin noted, was also born out of the rise of the great cities, their disturbingly heterogeneous crowds and newly scaled spaces demanding a point of reference that, while not refuting a certain instability, nevertheless served to dominate it aesthetically. Here the privileged point of view—of Hoffmann's observer keeping his careful distance from the marketplace, looking through "The Cousin's Corner Window" with opera glasses; of Poe and of Dickens watching the crowd; of Baudelaire losing himself in the swarming boulevards—attempted to preserve a sense of individual security that was only precariously sustained by the endless quest of the detective tracking his clues through the apparent chaos of modern urban life.[1]

In the context of the nineteenth-century city, the alienation of the individual expressed by writers from Rousseau to Baudelaire was gradually reinforced by the real economic and social estrangement experienced by the majority of its inhabitants. For Benjamin Constant, writing in the aftermath of the French Revolution and Napoleonic Empire, urban estrangement was a consequence of the centralization of the state and the concentration of political and cultural power, where all "local customs" and community bonds were brutally severed... For Marx, writing some thirty years later, individual estrangement had become class alienation...

This sense of estrangement was intellectually reinforced by the disturbingly transient qualities of the twin foundations of certainty for the nineteenth century—history and nature. The uncanny habit of history to repeat itself, to return at unexpected and unwanted moments; the stubborn resistance of nature to the assimilation of human attributes and its tragic propensity to inorganic isolation, seemed, for many, to confirm the impossibility of "living comfortably" in the world. Estrangement, in these terms, seemed a natural consequence of a conception of history, of the implacable impulsion of time that, while sweeping away the past in favor of the future, was necessarily uncertain only about the present. The remedies to such uncertainty, which ranged from revolution to restoration, from reform to utopia, were equally caught in the dilemmas of temporality, tied to the inhospitable context of the here-and-now at the same time as imagining a there-and-then. This anxiety of time, as expressed in the intellectual attempts to imagine impossible futures or return to equally impossible pasts, was accompanied by a fascination with the consequences of time's errors—the dystopian effects of unwonted interference with the natural developments of things, on the one hand, and the psychological effects of past and future shock on the other.

Gradually generalized as a condition of modern anxiety, an alienation linked to its individual and poetic origins in romanticism, the uncanny finally became public in the metropolis. As a sensation it was no longer easily confined to the bourgeois interior or relegated to the imaginary haunts of the mysterious and dangerous classes; it was seemingly as disrespectful of class boundaries as epidemics and plagues. Perhaps this is why, from the 1870s on, the metropolitan uncanny was increasingly conflated with metropolitan illness, a pathological condition that potentially afflicted the inhabitants of all great cities; a condition that had, through force of environment, escaped the overprotected domain of the short story. The uncanny here became identified with all the ...phobias associated with spatial fear...

For the modernist avant-gardes, the uncanny readily offered itself as an instrument of "defamiliarization" or *ostranenie*; as if a world estranged and distanced from its own nature could only be recalled to itself by shock, by the effects of things deliberately "made strange." Expressionist artists and writers from Kubìn to Kafka explored the less nostalgic conditions of the modern uncanny, pressing the themes of the double, the automat, and derealization into service as symptoms of posthistorical existence. Symbolists, futurists, dadaists, and of course surrealists and metaphysical artists found in the uncanny a state between dream and awakening particularly susceptible to exploitation. In this way, the uncanny was renewed as an aesthetic category, but now reconceived as the very sign of modernism's propensity for shock and disturbance.

"Estrangement from the world," noted Adorno, citing Freud's essay on the uncanny, "is a

2. Theodor Adorno, *Aesthetic Theory*, translated by C. Lenhardt, edited by Gretel Adorno and Rolf Tiedmann (London and New York: Routledge and Kegan Paul, 1984), 262.
3. Ibid.
4. Ernst Bloch, "Building in Empty Spaces" (1959), in *The Utopian Function of Art and Literature*, trans. Jack Zipes and Frank Mecklenburg (Cambridge: MIT Press, 1988), 186-199.

moment of art." For Adorno, indeed, the secret movement of the uncanny was the only way to explain why "the most extreme shocks and gestures of estrangement emanating from modern art—seismograms of a general, inescapable mode of response—are closer to us than past art which merely seems close because of its historical reification."[2] The artistic techniques of estrangement, he argued, rendered art less alienated than the condition they seek to address:

There is no denying that the antagonistic condition Marx called alienation was a power -ful leaven for modern art. However modern art was not simply a replica or reproduction of that condition but has denounced it in no uncertain terms, transposing it into an imago. *In so doing modern art became the opposite other of an alienated condition. The former was as free as the latter was unfree.*[3]

...As a concept, then, the uncanny has, not unnaturally, found its metaphorical home in architecture: first in the house, haunted or not, that pretends to afford the utmost security while opening itself to the secret intrusion of terror, and then in the city, where what was once walled and intimate, the confirmation of community... has been rendered strange by the spatial incursions of modernity. In both cases, of course, the "uncanny" is not a property of the space itself nor can it be provoked by any particular spatial conformation; it is, in its aesthetic dimension, a representation of a mental state of projection that precisely elides the boundaries of the real and the unreal in order to provoke a disturbing ambiguity, a slippage between waking and dreaming...

...The possibility that the theme of the uncanny, considered both historically and in its post-Freudian dimensions, opens questions that are larger than their simple illustration in architectural projects, questions that have stubbornly refused solution in politics as in design and that seem still pertinent to our late twentieth-century condition. In this sense, I want to use the different connotations of the theme both suggestively and critically, understanding its various textual and architectural manifestations are problematic contributions to a yet unfinished history that pits the homely, the domestic, the nostalgic, against their ever-threatening, always invading, and often subversive "opposites."

For if the theoretical elaboration of the uncanny helps us to interpret the conditions of modern estrangement, the special characteristics of architecture and urbanism as arts of spatial definition allow us to advance the argument into the domain of the tangible. Here it is that the "void" described by *posthistoire* philosophy is almost uncannily repeated in the world, that the question of the "unhomely home" finds its most poignant expressions and equally troubling solutions. This occurs on a number of interrelated levels, both literal and phenomenal. On the literal plane, the "empty spaces" appropriated or created by urbanism—the clearing of vacant or occupied territory—are paralleled on the phenomenal plane by the tabula rasa imagined by modernist utopias, to the point where both levels intersect in the commonplaces of modern urban development. The task of filling these voids—what Ernst Bloch has termed the "hollow spaces of capitalism"—is given over to architecture, which is forced, in the absence of a lived past, to search for posthistorical grounds on which to base an "authentic" home for society.[4] Thus, on an even more literal level, architecture finds itself "repeating" history, whether in traditional or avant-garde guise, in a way that itself gives rise to an uncanny sense of déjà vu that parallels Freud's own description of the uncanny as linked to the "compulsion to repeat."

The apparently irreconcilable demands for the absolute negation of the past and full "restoration" of the past here meet in their inevitable reliance on a language of architectural forms that seem, on the surface at least, to echo already used-up motifs *en abîme*.

Deployed in this way, the uncanny might regain a political connotation as the very condition of contemporary haunting; what in the sixties was so overtly a presence in theory and practice, a presence that largely denied the formal in architecture in favor of social practice, utopian or material, is now, in the nineties, apparently suppressed by an ostensibly nihilistic and self-gratifying formalism... In contemporary architecture, the incessant reference to avant-garde techniques devoid of their originating ideological impulse, the appearance of a fulfilled aesthetic revolution stripped of its promise of social redemption, at least approximates the conditions that, in Freud's estimation, are ripe for uncanny sensations.

Anthony Vidler (b. July 4, 1941) is Dean and Professor of the Irwin S. Chanin School of Architecture at The Cooper Union, New York. He is the author of *Warped Space: Art, Architecture, and Anxiety in Modern Culture* (2000), and *The Architectural Uncanny: Essays in the Modern Unhomely* (1992), both published by The MIT Press, and other books.

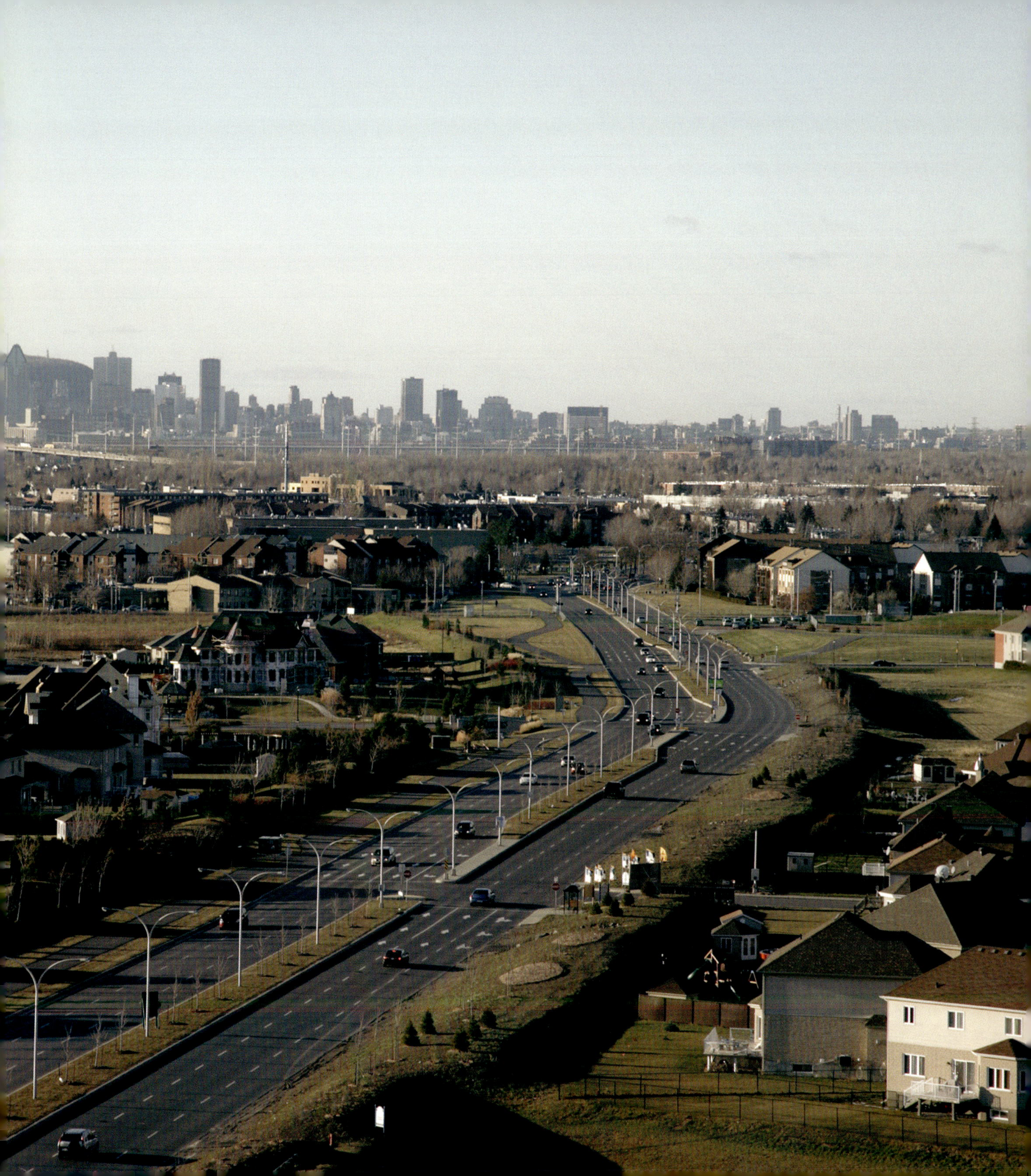

Artworks & Biographies

All descriptive texts and biographies are co-written by Amanda Cachia and the artists.

Isabelle Hayeur, still from *Losing Ground*, 2009.

BGL

OPPOSITE
***Pinocchio*, 2009**
Chainsaw, clamp, plywood, vinyl, paint; with pair of speaker fans (fans, wood, plastic) 112 x 115 x 126 inches; each fan: 16 x 32 x 19 inches
Courtesy of the artists and Diaz Contemporary, Toronto

Quebec-based collective BGL includes the members Jasmin Bilodeau, Sébastien Giguère and Nicolas Laverdière. Their contribution to *Ecotopia* is unique in that their sculptural installation, *Pinocchio*, constructed from machine-cut plywood for the tree trunk and branches and plastic for the green leaves, is also haptic. Several speaker fan boxes attached to the walls around the sculpture blow wind onto a large tree. The tree slowly turns, occasionally stops, and then starts again. An emblematic chainsaw balances precariously on one of the branches, and while visitors might hesitate getting close to the moving branches for fear of being lopped off, the artists have included a small gallery bench close to the tree, so that visitors are encouraged to sit underneath the foliage to observe the mesmerizing movement of this tree in mimicry. Nearby, fallen branches and leaves are scattered across the gallery ground as a reminder of the tree's fragility. Writer Carol-Ann Ryan has effectively discussed BGL's choice of title for this work, referencing the fairytale character Pinocchio—a marionette who wanted to become a real boy but whose nose would grow when he told lies. She says, "BGL's Pinocchio will never grow and, perhaps, its dangling chainsaw represents the reality of clear-cutting in Canada's forests. The continuation of this practice requires a thoughtful balance between conservation and sustainability, which seems reflected in the delicate assembly that keeps BGL's sculpture upright."[1]

BGL (Jasmin Bilodeau, b. 1973 in Lac-Mègantic, Quebec; Sébastien Giguère, b. 1972 in Arthabaska, Quebec; and Nicolas Laverdière, b. 1972 in Montreal, Quebec) is a Quebec City-based trio who met while studying at Laval University. For over a decade, BGL has created sculptures, prints and installations that critique the commercialism of contemporary culture, often using humour. Having also explored larger themes such as death, truth, and the role of art, BGL aims to make the viewer fascinated and uncomfortable at the same time. The trio playfully explores the contradictions in the material and the natural world. Nature collides with the manufactured, disrupting predictable narratives and questioning the values of society. BGL was included in *Sous les ponts, le long de la rivière II* at Casino Luxembourg, 2005; the 9th Havana Biennial, 2006; La Biennale de Montréal, 2007; and la Bienal del fin del Mundo in Ushuaia, 2007. Recent group exhibitions include *On Being an Exhibition* at Artists Space in New York, 2007; *Caught in the Act: The Viewer as Performer* at the National Gallery of Canada, 2008–09; *Manoeuvres/Maniobres* at Galeria Toni Tàpies in Barcelona, 2009; Nuit Blanche Paris, 2011; Nuit Blanche Calgary, 2012; and *Oh, Canada* at the Massachusetts Museum of Contemporary Art (MASS MoCA), 2012. They are represented by Parisian Laundry in Montreal and Diaz Contemporary in Toronto.

1. Carol-Ann Ryan, "BGL: New Sellutions," *Magenta* magazine online, Winter 2010, http://www.magentamagazine.com/2/reviews/bgl-new-sellutions (accessed August 2012).

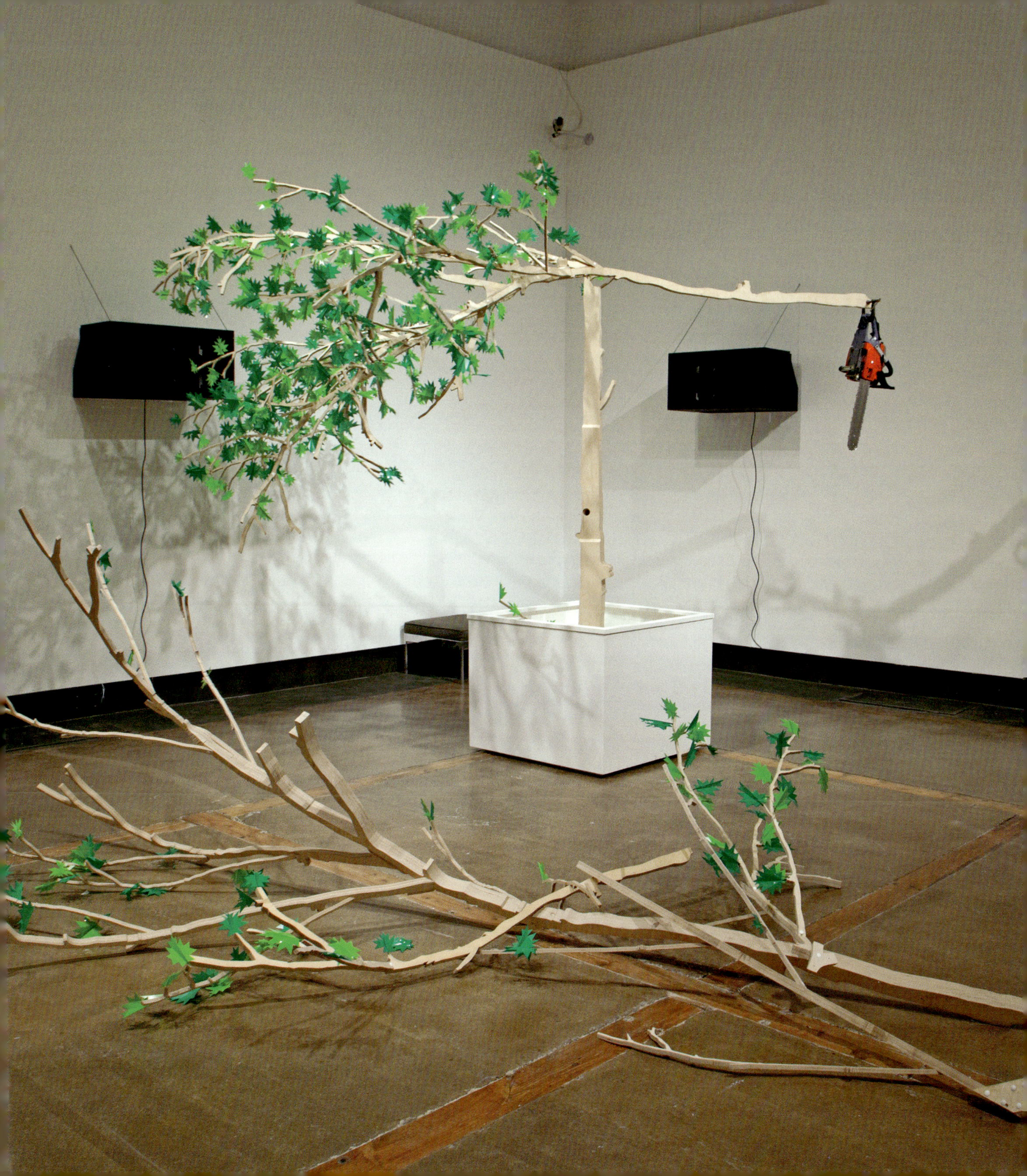

David Brooks

OPPOSITE
***Unfinished Section of Sidewalk with Two Palms and Rebar*, 2012**
Concrete, rebar, plant material, 30 x 22 x 36 inches. Courtesy of the artist and American Contemporary, New York

Using the material language of sidewalks, cranes, roofs, steps, plant material and infrastructural devices, Brooks addresses the irrational efforts we exert to maintain a conflicted relationship with the natural world, as evinced in the common traits of our built environment. In *Unfinished Section of Sidewalk with Two Palms and Rebar*, a slab of concrete is tipped on its side with two neglected palms protruding from it, and supporting it upright. This object could be a material artifact and commonplace sight in areas of extreme suburban sprawl where massive housing construction projects remain unfinished due to the fluctuations in the housing market. These are the frontlines where nature and culture are in daily confrontation over a new form of normalcy.

David Brooks (b. 1975 in Brazil, Indiana) lives and works in New York. His work considers the relationship between the individual and the built and natural environment, and investigates how cultural concerns cannot be divorced from the natural world. Brooks received his BFA from Cooper Union and an MFA from Columbia University. His work has been exhibited nationally and internationally at the Changwon Sculpure Biennale in South Korea, Galerie für Landschaftskunst in Hamburg, Miami Art Museum, Dallas Contemporary, Nouveau Musée National de Monaco, Bold Tendencies in London, as well as Gavin Brown's Enterprise, Tanya Bonakdar Gallery, Marlborough Chelsea, and the Sculpture Center in New York City. He also had a major two-year installation at MoMA PS1 in New York in 2010. Recent commissions include *Desert Rooftops* in Times Square, New York, a 5000-square-foot earthwork commissioned by the Art Production Fund (2011); and a large-scale project for The Cass Sculpture Foundation, UK (2012). Brooks has also been the recipient of numerous awards, including a Socrates Sculpture Park Fellowship (2009), a research grant to the Peruvian Amazon from the Coypu Foundation (2009), a grant from the Foundation for Contemporary Arts (2010), and a research grant to the Bolivian Amazon from the Explorers Club (2011).

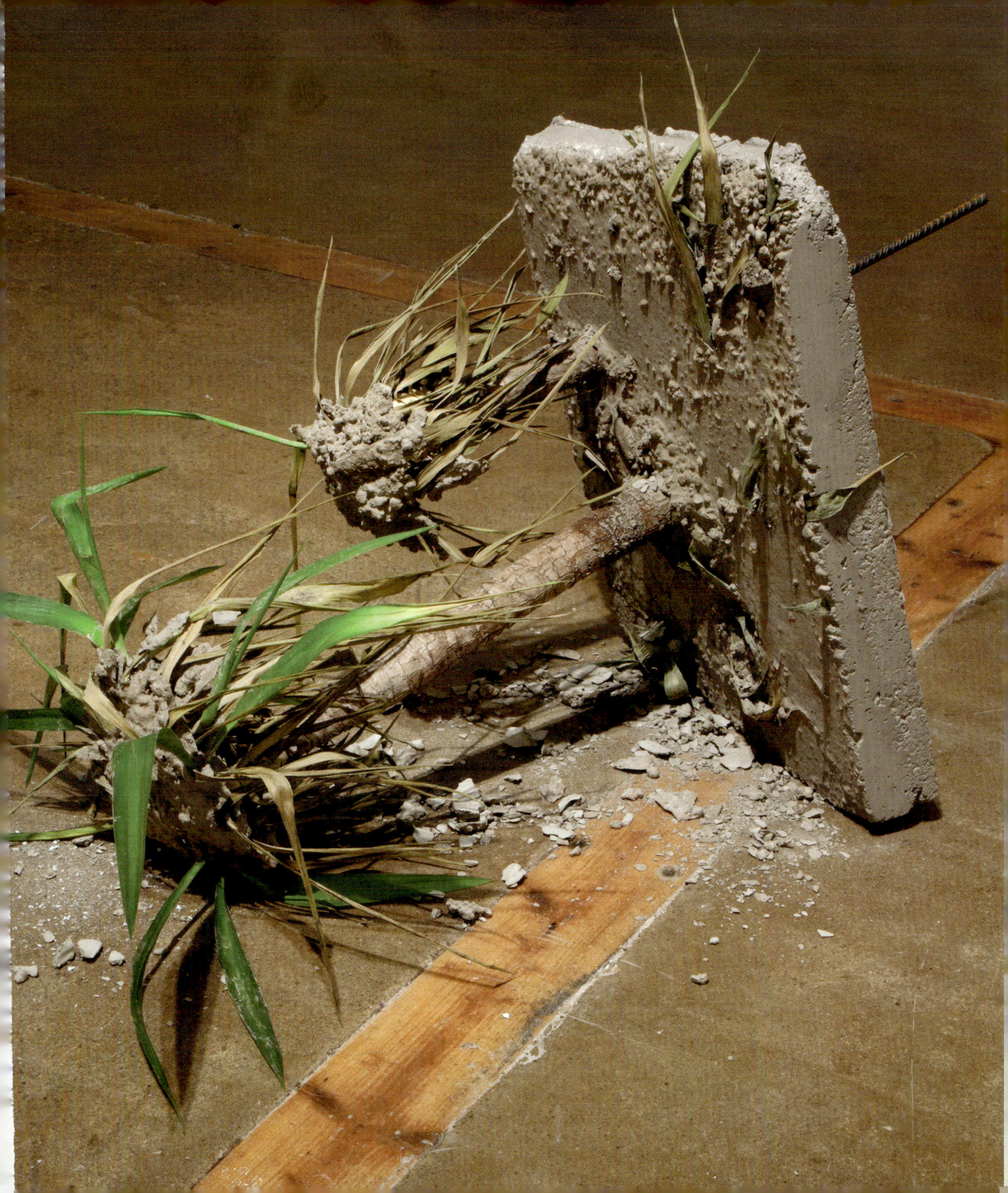

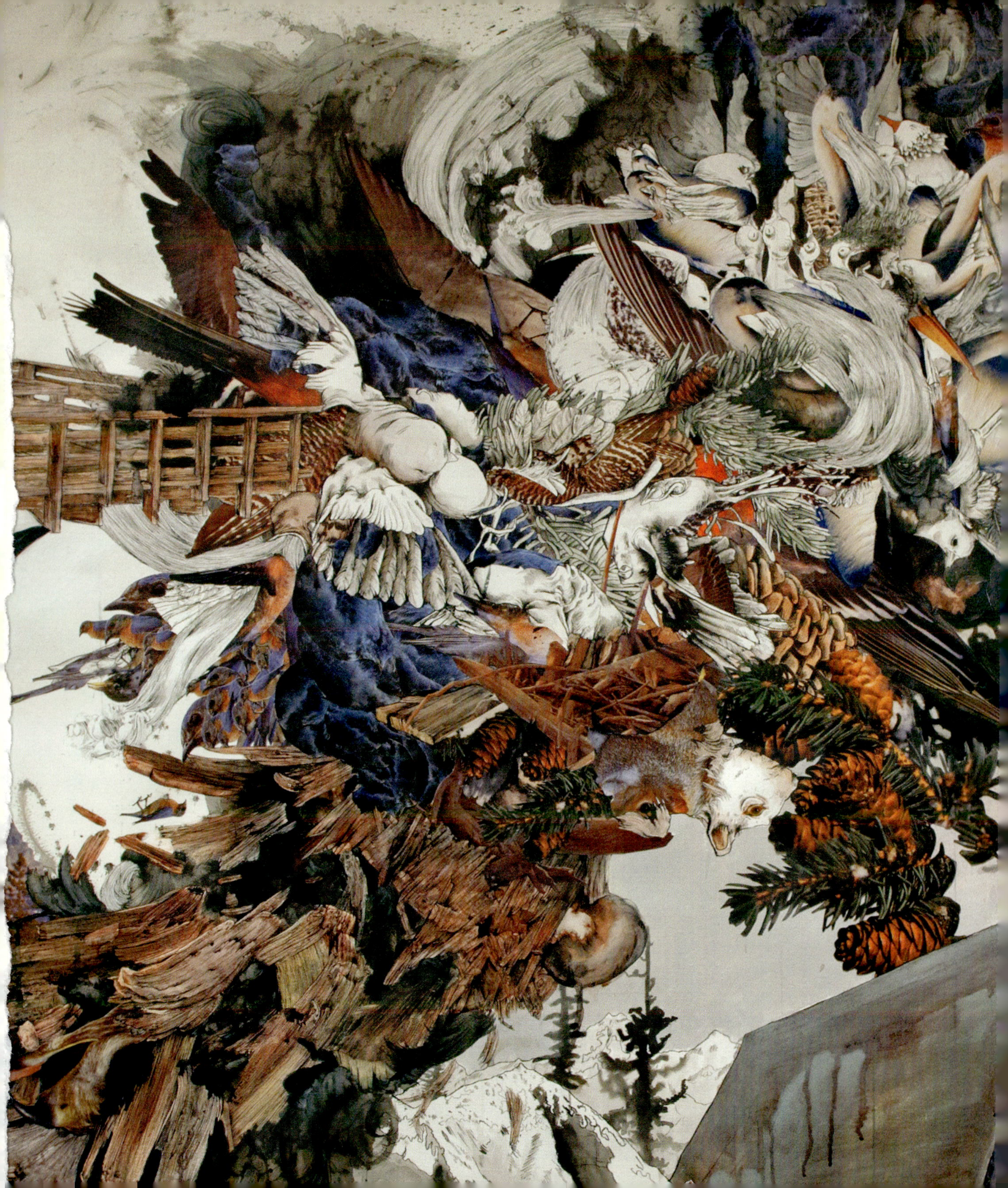

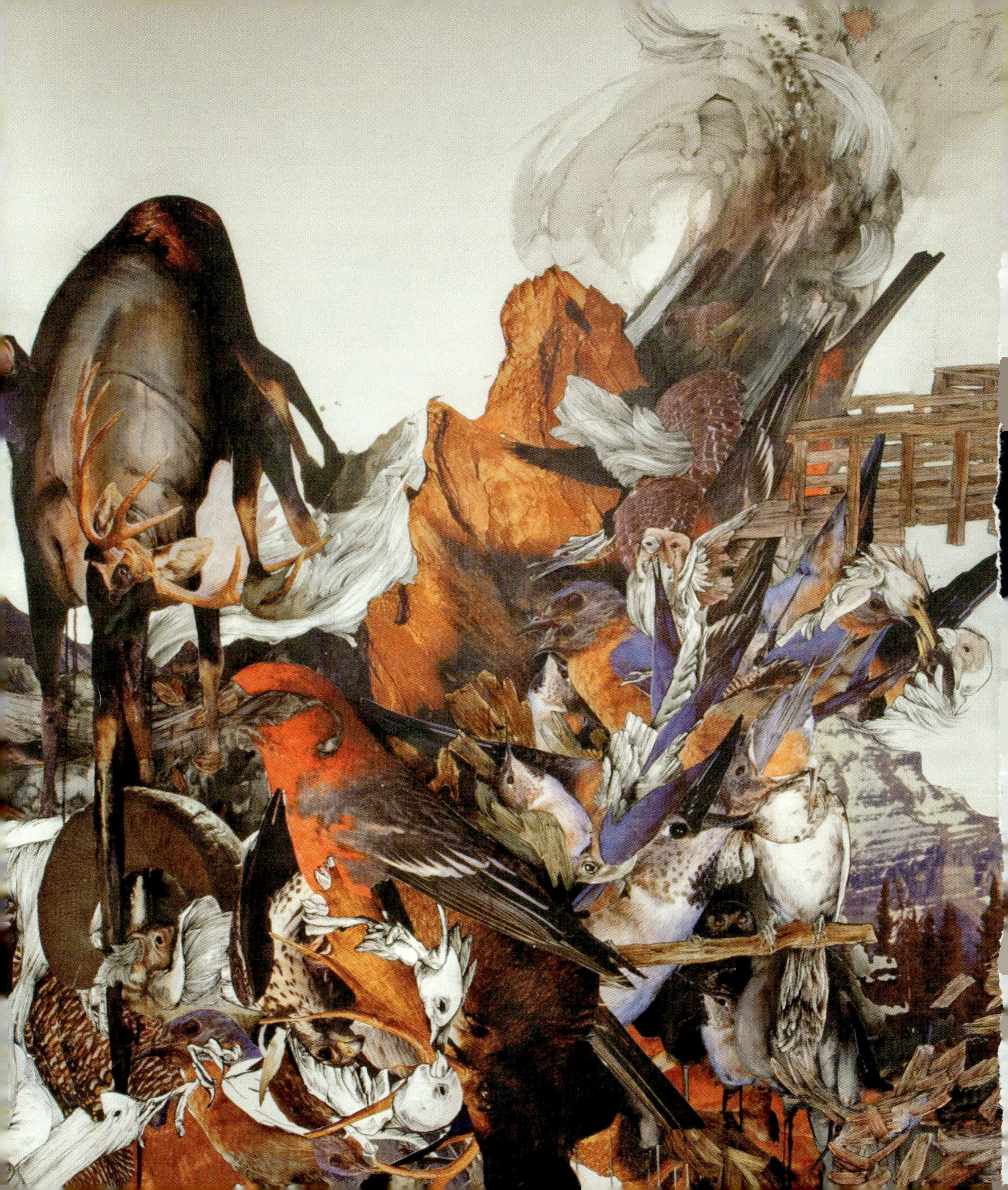

Dagmara Genda

OPPOSITE and FOLLOWING PAGES
***Panorama*, 2012**
Collage with ink, latex, and acrylic on paper, 52.5 x 301.4 inches
Courtesy of the artist

The artist wishes to thank the Saskatchewan Arts Board for their assistance in making this project for *Ecotopia* possible.

Panorama is a 360˚ paper mural of Canada displayed in traditional panoramic fashion, in a cylinder hung from the ceiling into which people can enter. The collaged images come from nature books, travel brochures and other somewhat superficial representations of the Canadian landscape. *Panorama* refers to signs and regulations in national parks and other areas where people come to enjoy Canada's wilderness. It marks the government's efforts to protect its natural spaces while, ironically, changing and commodifying them through that very protection.

Genda is interested in the way that Canada markets its identity to the world, and the "ownership" of wilderness and how that has become a cultural commodity. This is part of Canadian heritage as exemplified by the Group of Seven, who painted scenes of Canadian wilderness partly in order to promote a homogenous image of Canadian nationality. This form of myth-building informs how our tourism industry markets Canada and how many books on Canadian wilderness are written and illustrated. While no longer part of an explicit campaign of nation building given that discourses on colonialism, multiculturalism and immigration have become more prevalent, such romantic images still possess a strong hold on the collective imagination.

Dagmara Genda (b. 1981 in Koszalin, Poland) works primarily in drawing and drawing-based installation. Genda emigrated from Poland to Canada in 1985 and continued to move from place to place across the country as well as Alaska and London, UK. Her frequent moves have led to an interest in regional identities, art histories, and architectures. Her work can be divided into two primary streams. One prominently features varied representations of the Canadian landscape, from Group of Seven paintings to Canadiana coffee table book photography, and the other focuses on Polish architecture, specifically the Palace of Culture and Science in Warsaw and mass-produced apartment blocks. Her work has been exhibited nationally from coast to coast, including the Mendel Art Gallery in Saskatoon, Saskatchewan; and the Walter Phillips Gallery in Banff, Alberta, among others. She has also exhibited internationally in New York and North Carolina. Genda was the winner of 3rd Ward's 2011 Winter Open Call, for which she received a 3-month residency at 3rd Ward's Brooklyn location. In 2011, she also received a commission to create a permanent vinyl installation in Chelsea, New York. She is also the recipient of numerous other awards, including Saskatchewan Arts Board Grants, 3rd place at Durham's 56th Annual Juried Show (Durham, North Carolina), and the 2009 Best Drawing Award at the Toronto Outdoor Art Exhibition. She holds an Honours BFA from the University of Manitoba, an MFA from the University of Western Ontario and an MA in Cultural Studies from Birkbeck College, University of London. She currently lives in Saskatoon.

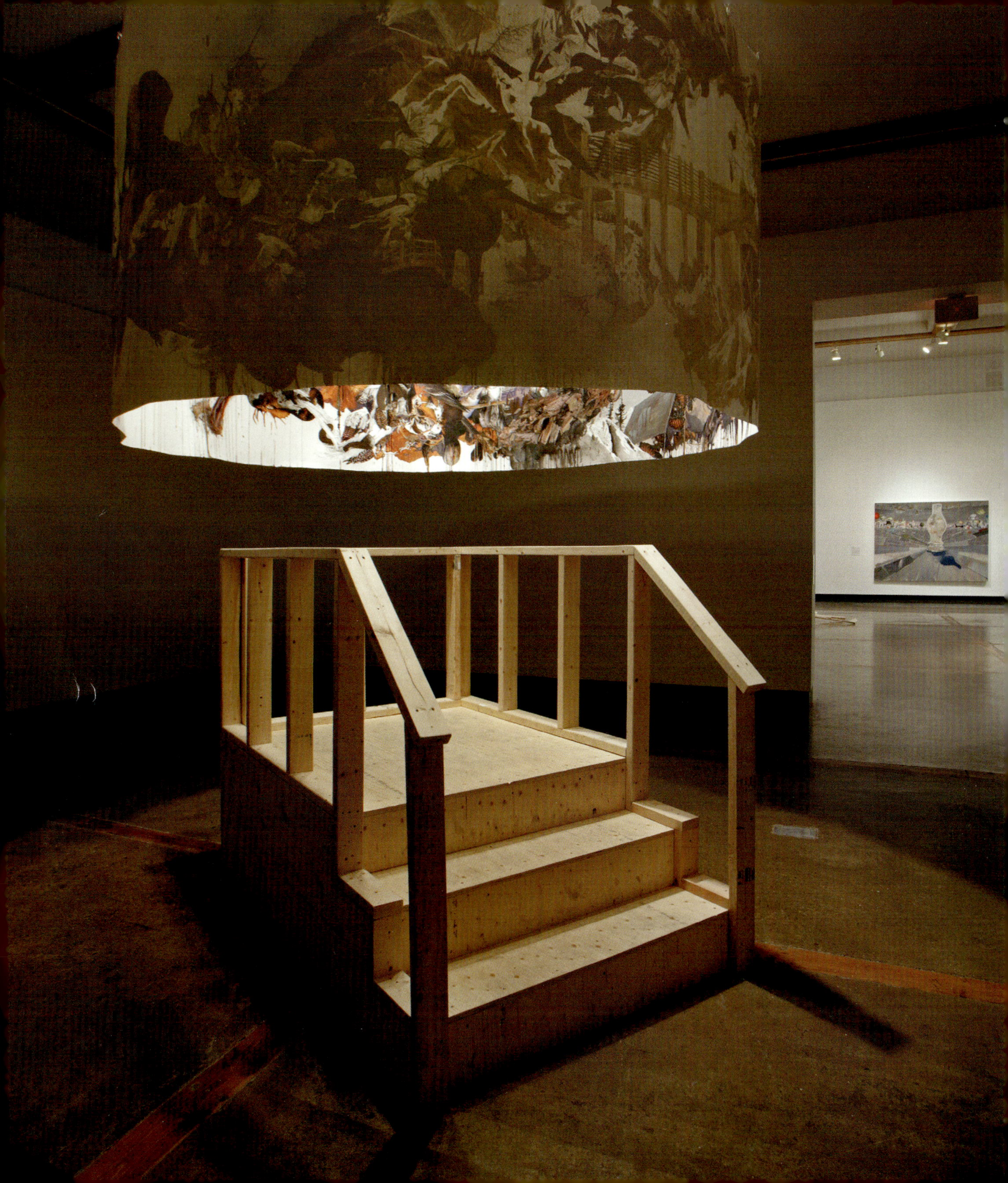

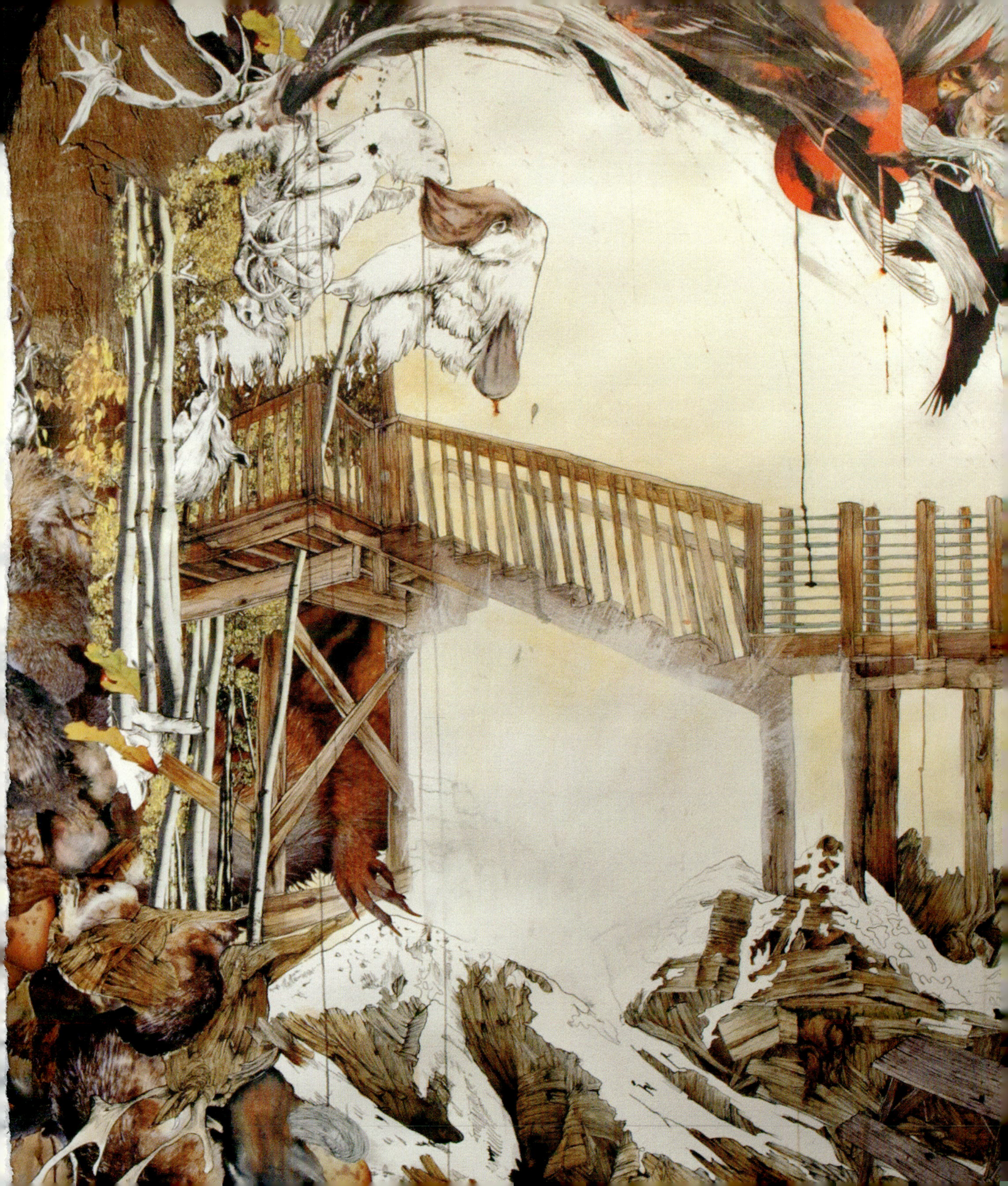

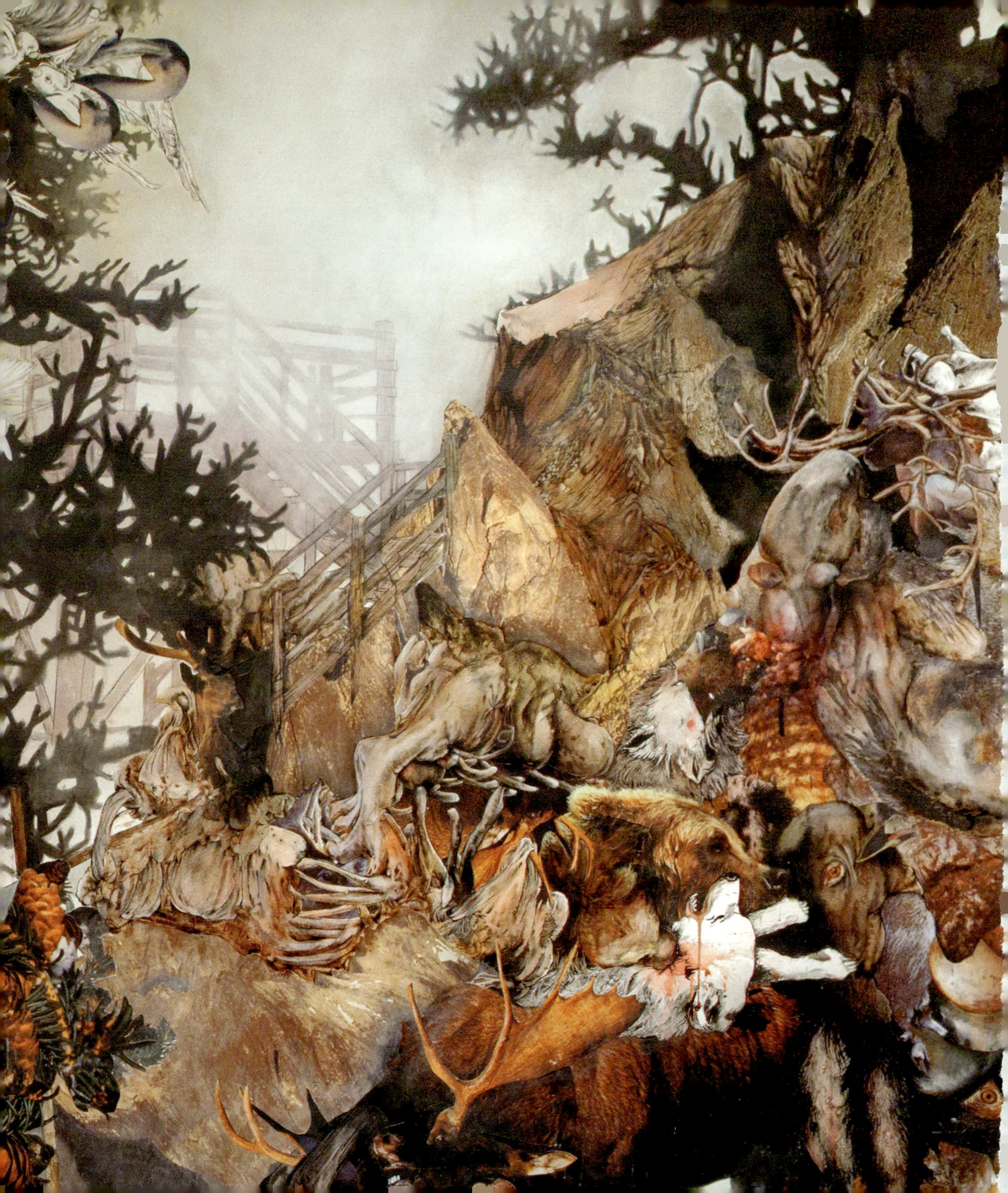

Rodney Graham

BELOW and OPPOSITE
***Ponderosa Pines, Princeton B.C./ CAT hi-way yellow*, 1992**
Edition 5/15, 5 chromogenic prints
Each 10 x 8 inches
Collection of McCarthy Tétrault, LLP, Toronto

Ponderosa Pines, Princeton B.C./ CAT hi-way yellow are portraits of individual trees broken up by yellow monochrome panels. In a British Columbian context, the trees relate to the ongoing level of anxiety over clear-cutting. That some images are of trees in Vancouver's Stanley Park marks this directly. Writer and academic Sarah Parsons says, "Trees serve as a rich subject because they carry such a complex web of meanings and metaphors (language, life, etc.). In formal terms, they echo Graham's interest in looping and repetition: they repeat themselves as branches grow to resemble roots, an aspect he highlights by inverting them. After September 11, 2001, one of his inverted trees in a group show was described by Hilarie M. Sheets in *ARTnews* as "a metaphor for the nostalgic sepia-toned world that has now been turned upside down."[1] This work also suggests the extent to which our experience of landscape, so crucial to Canadian culture, is mediated rather than "natural." How do we learn to see what is "natural" or "artificial"? What are we editing out and how does that action affect our consciousness about the environment? Parsons states: "Graham's inverted tree photographs have become somewhat of a trademark, visually appealing in their majestic clarity, but still baffling to many viewers."[2]

Rodney Graham (b. 1949 in Abbotsford, British Columbia) studied art history at the University of British Columbia from 1968 to 1971 and at Simon Fraser University in Vancouver from 1978 to 1979. An artist and musician, Graham has created a varied body of work since the 1970s, making use of such diverse media as film, photography, installation, painting, prints, literature and music. He has visualized the interconnections between the various layers of everyday life and cultural history with a great deal of poetic irony, finesse and conceptual originality. Exploring and reflecting on works of art, music, and literature, as well as events in the intellectual history of all periods, he has created a complex œuvre that delicately balances reality and fiction. Graham has exhibited widely throughout North America and Europe for over twenty years and, in 1997, he represented Canada at the Venice Biennale. In 2005 he was the subject of a major survey exhibition at the Vancouver Art Gallery that toured to the Museum of Contemporary Art, Los Angeles, among other venues in North America. Graham's work can be seen in the collections of the Museum of Modern Art and the Metropolitan Museum of Art in New York, and the Centre Pompidou in Paris, among others. He is represented by Lisson Gallery, London; Donald Young Gallery, Chicago; 303 Gallery, New York; and Hauser & Wirth, Zurich.

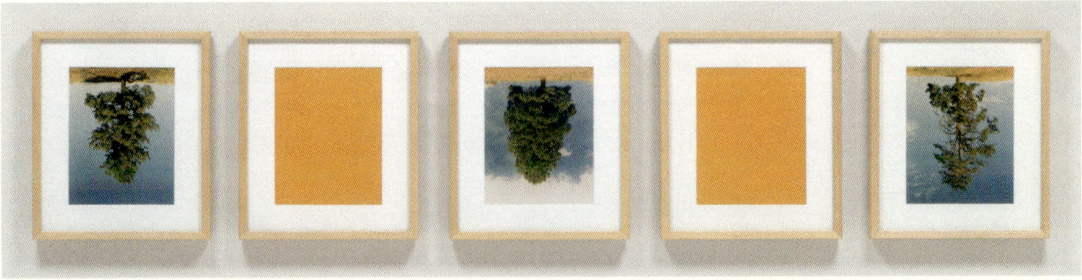

1. Sarah Parsons, "Rodney Graham: A Little Thought," *Canadian Art*, Spring 2004, http://208.70.246.208/art/features/2004/06/08/217/ (accessed August 2012).
2. Ibid.

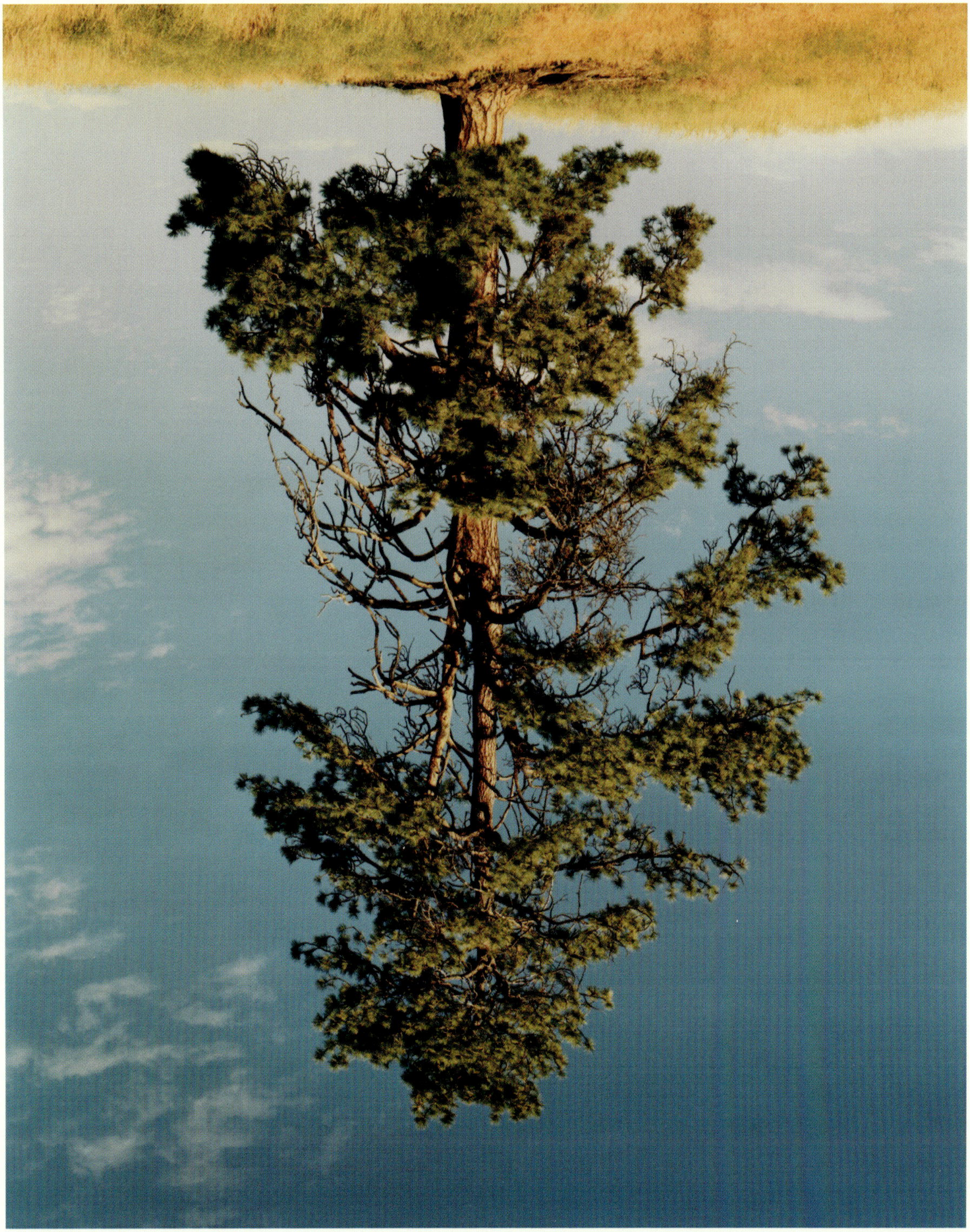

Isabelle Hayeur

OPPOSITE
***Ascendance*, 2005–2008**
Inkjet on polyester, 125 x 39 inches. Courtesy of the artist and Galerie Division, Montreal

FOLLOWING PAGES
***Losing Ground*, 2009**
High-definition video, 13:00 Courtesy of the artist and Galerie Division, Montreal

Ascendance is one of the photographs from Hayeur's *Excavations* series executed from 2005 to 2008. The term "excavation" can refer to work in construction and road building, as well as mining and archaeology. The montages result from a union of landscapes that have opposite or contradictory significance. The artist worked with conservation sites rich in natural and human history, then with disturbed sites and their forms of disappearance. Hayeur's compositions combine the homogeneity of new housing developments with UNESCO (United Nations Educational, Scientific, and Cultural Organizations) World Heritage sites. The artist also combines fossiliferous sites with various landscapes shaped by economic needs, such as garbage dumps and mines. These landscapes blend naturally; their disturbed and stripped aspects make them similar. In *Ascendance*, the shift from a monumental scale to a much smaller scale disturbs hierarchy: that which seemed immense and immutable becomes vulnerable and that which was insignificant becomes important. A lone, bare tree trunk and empty branches rest precariously on a large pile of soil and seemingly out of place in this new man-made nature.

Losing Ground was filmed in Quartier DIX30 in Brossard, the biggest and first lifestyle centre in Canada, and the video probes recently man-made territories so as to decipher humanity's relationships with the environment. Quartier DIX30 was designed to emulate an urban or downtown shopping experience for suburban dwellers living on the South Shore of Montreal. As well as shops and department stores, Quartier DIX30 includes live theatre, office towers, a hotel, a medical clinic, a toy superstore, and a spacious heated underground parking lot with 2000 free spaces. Hayeur critiques such urban sprawl and the resulting erosion and homogenization of rural environments. With its negation of city history, of geographic particularities, and thus of cultural memory, this standardized urbanization imposes its jarring presence in nature.

Isabelle Hayeur (b. 1969 in Montreal, Quebec) lives and works in Montreal. She is known for her large digital photomontages, videos and site-specific installations. An ecological, social and urban planning critique provides the framework for her artistic practice. Having lived in a suburb for some twenty years, she has been confronted with the sad spectacle of urban sprawl and all that it consigns to oblivion. Her approach is tied to this experience and draws from discourses about environmental issues and land use planning. She is particularly interested in feelings of alienation and dislocation. Her artworks have been shown in the context of numerous exhibitions and festivals, including those held at the National Gallery of Canada, the Musée d'art contemporain de Montréal, the Massachusetts Museum of Contemporary Arts (MASS MoCA), the Neuer Berliner Kunstverein in Berlin, the Tampa Museum of Art, the Museum of Contemporary Photography in Chicago, and Bruce Silverstein Gallery in New York. She is represented by Galerie Division in Montreal.

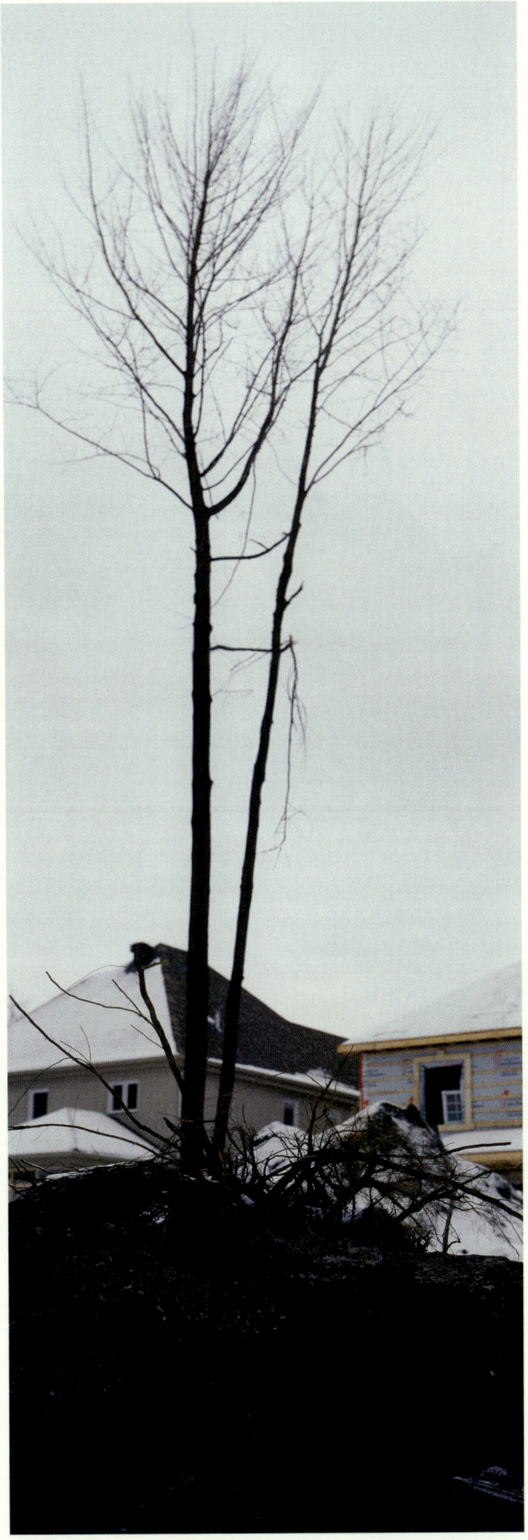

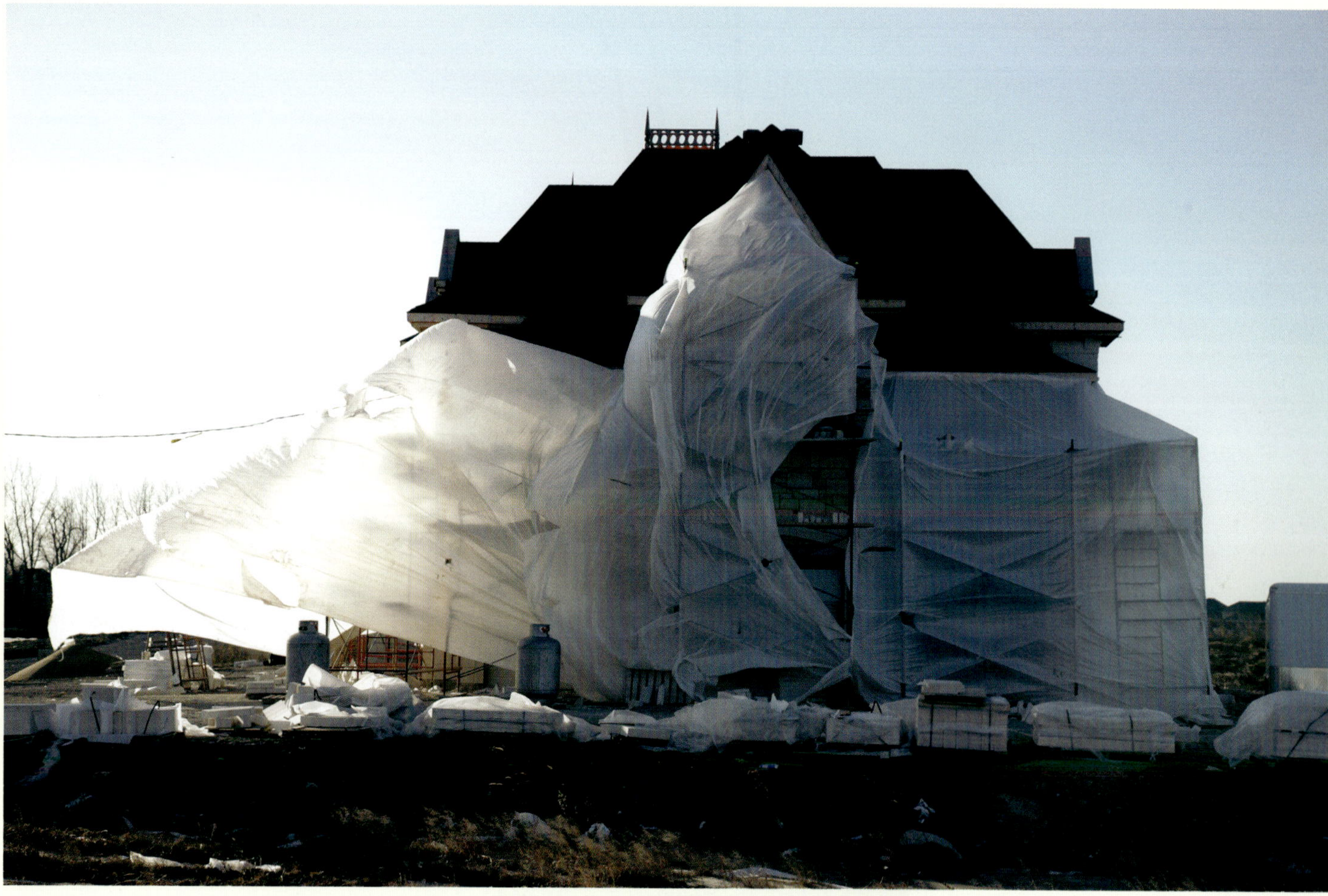

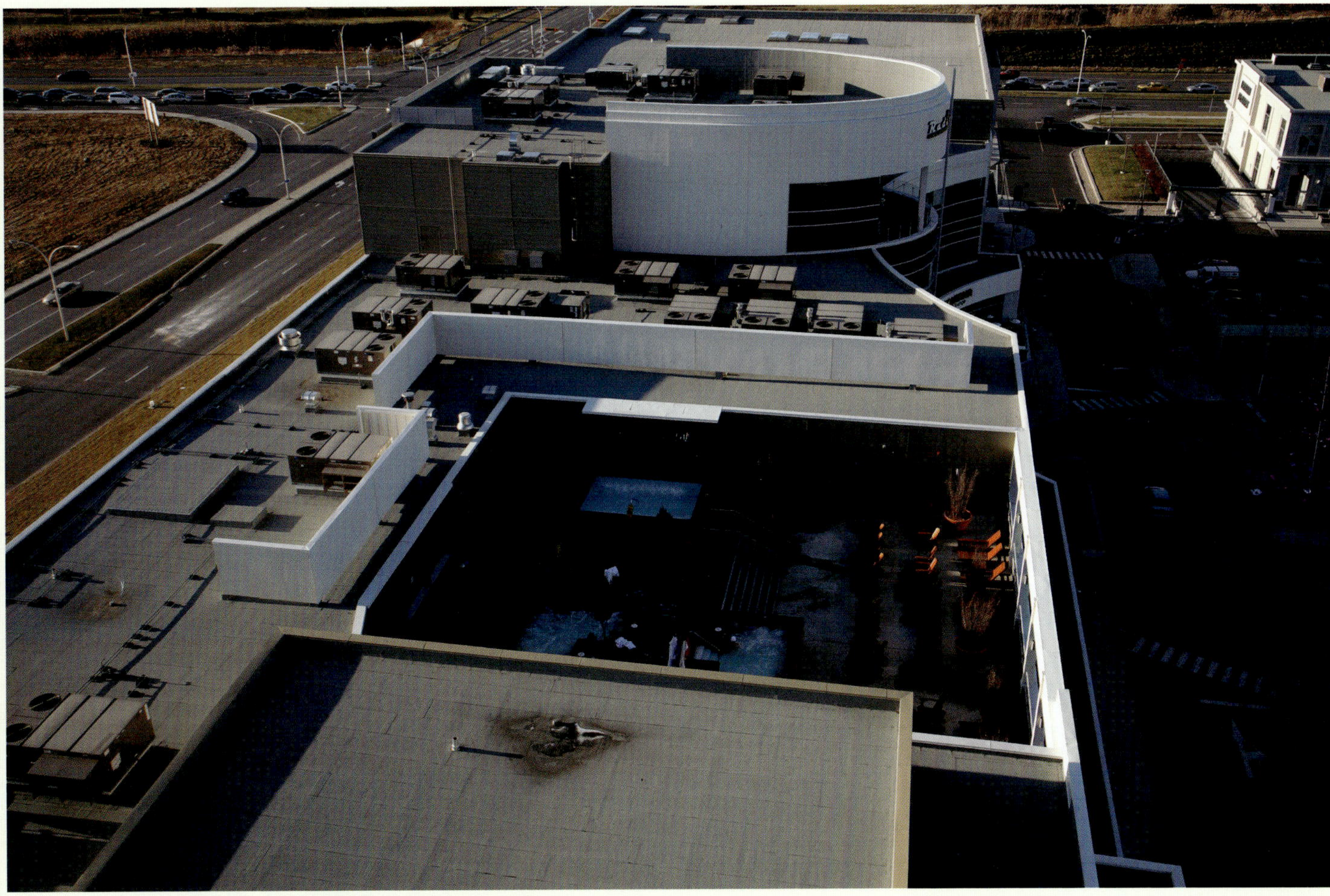

Tristram Lansdowne

BELOW and OPPOSITE
***Envelope*, 2012**
Watercolour on paper
36 x 25.5 inches
Courtesy of the artist

FOLLOWING PAGES
***The Nurse*, 2011**
Watercolour on paper
40 x 30.5 inches
Collection of Mark Zadorozny, Toronto

Lansdowne's work explores themes of function and permanence in relation to architecture and landscape through meticulously painted watercolours. Drawing from disparate systems of visual order such as nineteenth century Romantic landscape painting, 1960s utopian architecture, and scientific infographics, Lansdowne's work is focused on examining the artifice inherent in such idealized constructs. By combining specific, anecdotal visual description with intentionally broad art historical motifs and architectural propositions, the ideal is always infected by the vernacular and vice versa, giving way to a haunted landscape of conflicting visions.

The Nurse and *Envelope* both present landscapes dominated by abandoned structures that have become hosts for new architectural growth and flora. This new growth is, in fact, also retrograde, their forms adapted from mid-twentieth century architectural proposals designed to alleviate spatial and social issues. Part development proposal and part historical mash-up, these paintings suggest that the salvaging of both materials and principles might be viable options for the future. Lansdowne challenges us to reconsider our definitions of progress.

Tristram Lansdowne (b. 1983 in Victoria, British Columbia) currently lives in Toronto. He studied visual arts at the University of Victoria and at the Ontario College of Art & Design, where he received a Bachelor of Fine Arts in 2007. Lansdowne's work has been exhibited across Canada and the U.S. Recent exhibitions include *Fata Morgana* at LE Gallery, Toronto; *Empire of Dreams* at the Museum of Contemporary Canadian Art, Toronto; *Contained* at the Boston Center for the Arts; and *60 Painters* at Humber College, Toronto. He was also chosen as a semi-finalist in the 2011 RBC Canadian Painting Competition.

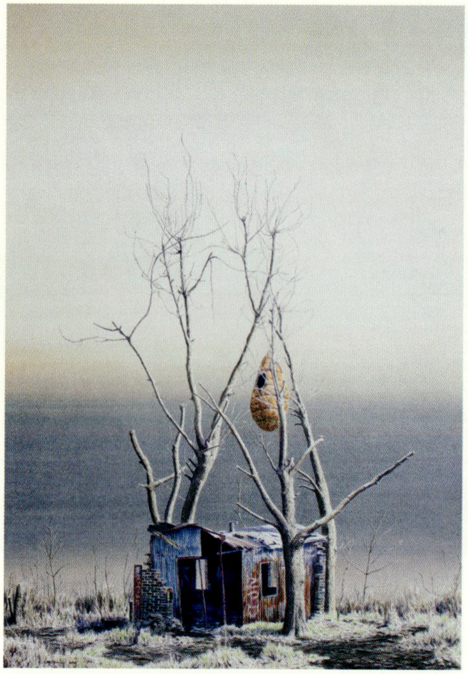

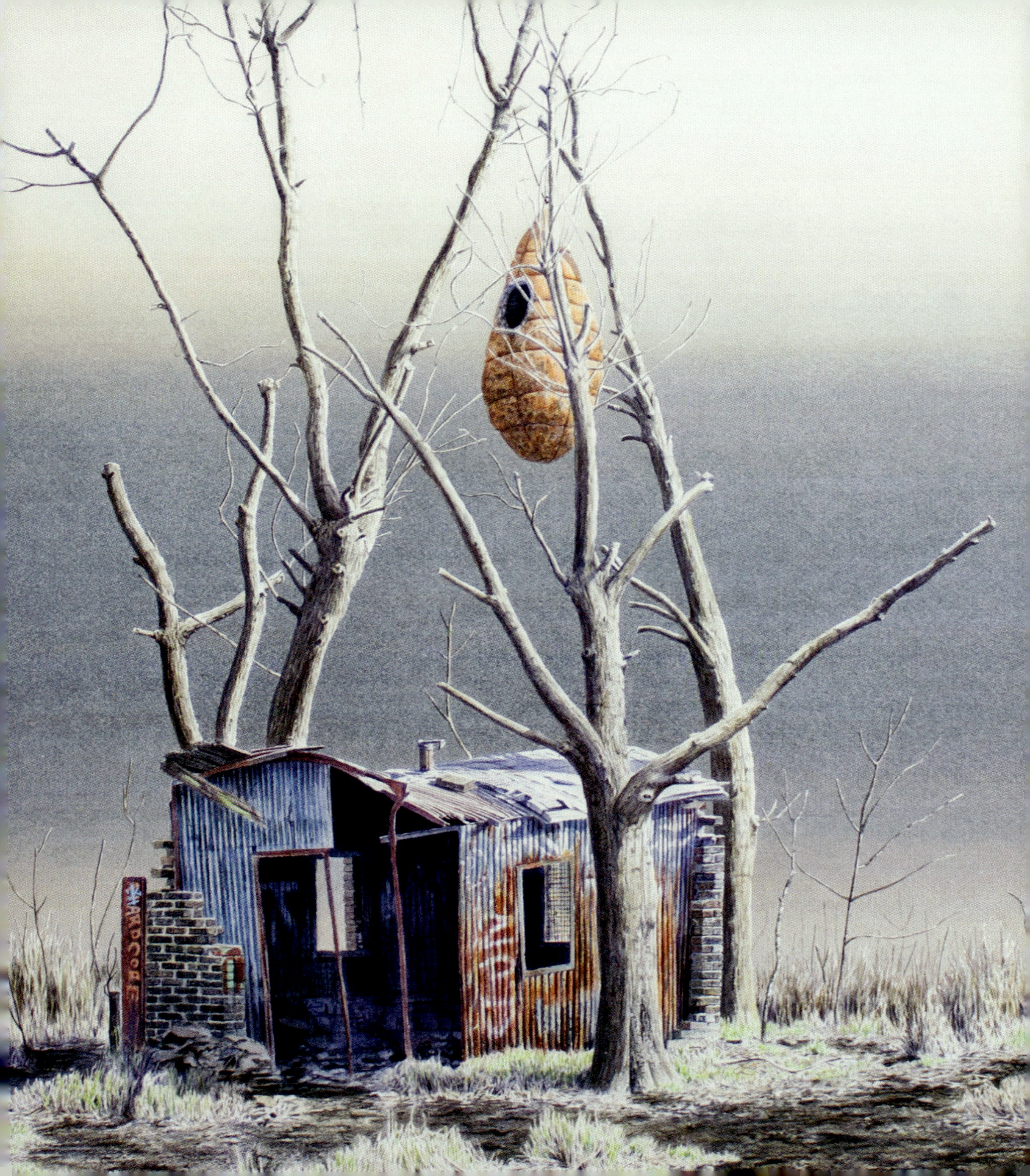

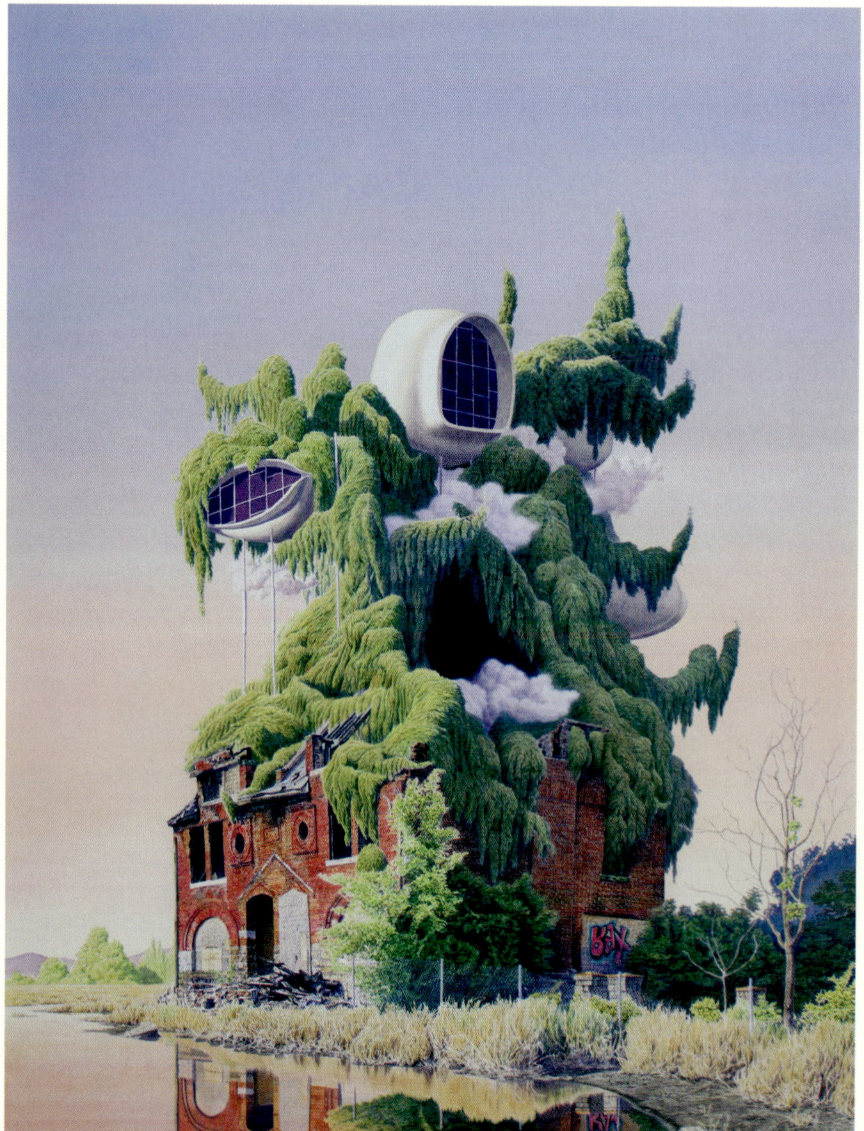

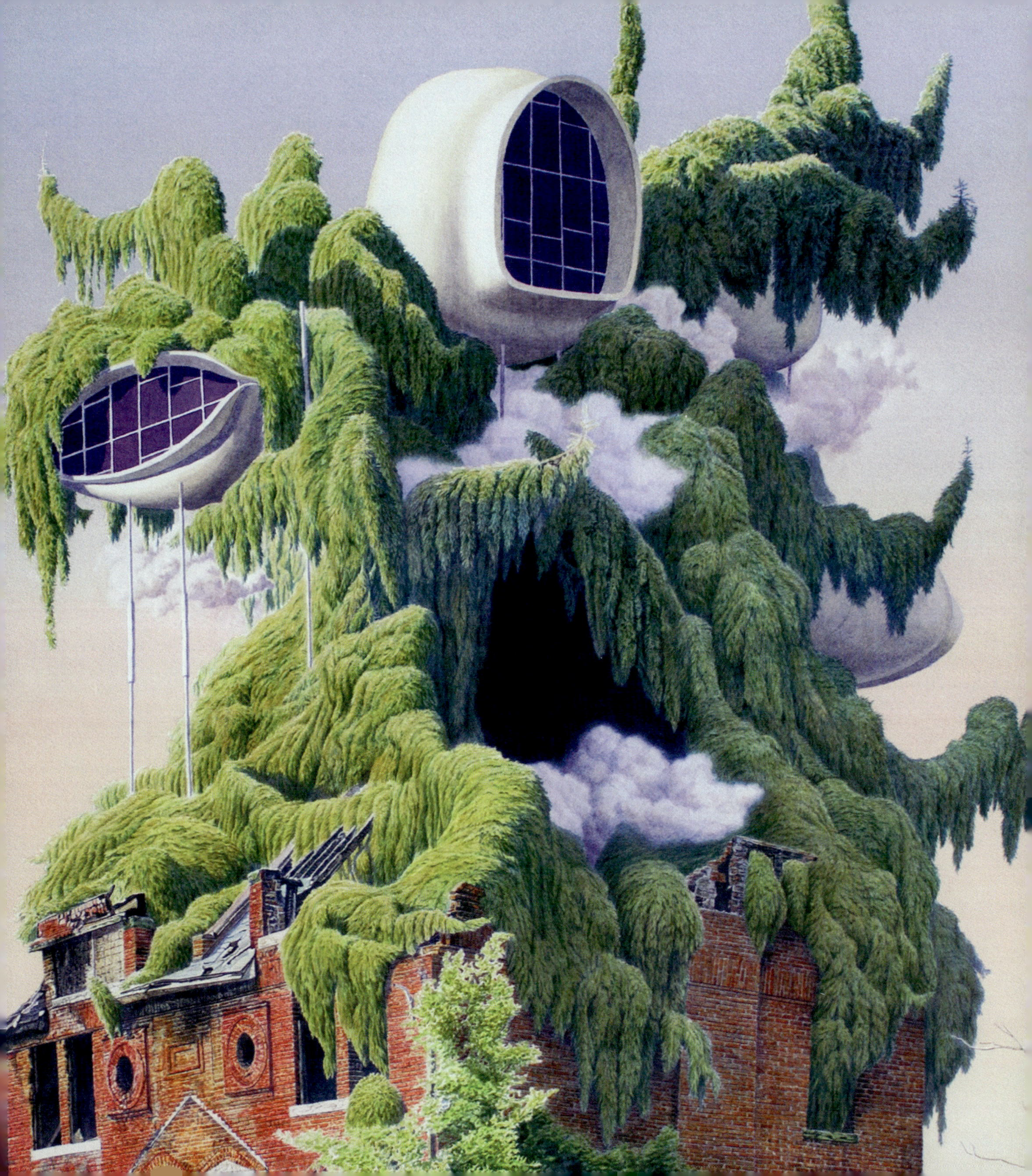

Maude Léonard-Contant

BELOW (from left)
***Cacti*, 2012**
Styrofoam, playdough, digital prints, flocking, wire
Dimensions variable

***Creeper*, 2012**
Coloured crayon, mdf, plexiglas, polymer
49.6 x 38.2 x 3.9 inches

***Coco-fesse*, 2011**
Rubber foam, ink
7.9 x 11.8 x 12.6 inches

Courtesy of the artist

Léonard-Contant explores strategies of simulation such as trompe l'oeil, mise-en-abîme and camouflage in these pieces, blurring the boundaries between fake and real, between interior design and landscape architecture. This work also grapples with themes of decay, illusion and passage from interior to exterior. While presenting an obvious physicality, they also seem to fade away, or to metamorphose themselves: cubes of marble are actually made from playdough, while what looks like concrete is mere papier mâché. The juxtaposition of the objects creates a sibylline dialogue between each of them: garden plants want to become enlarged cubist trinkets while swarms of butterflies transform into images of hysterical animals. As an ensemble, they interact as if they had chosen to hang out together in order to invent a story whose plot remains obscure. Proud protagonists of an ambiguous fiction, they invite viewers to indulge in multiple readings and interpretations.

Maude Léonard-Contant (b. 1979 in Joliette, Quebec) is a Canadian artist currently based in Switzerland. Working primarily in sculpture, her thoughtful handling of materials allows the objects and atmospheres she sets up to deploy cryptic fictions. She is fond of reminiscence and mirages. She has exhibited her work at various institutions, including Stadtgalerie, Bern; Abrons Art Center, New York; Kala Art Institute, Berkeley; Market Gallery, Glasgow; Centre Clark, Montreal; and Musée d'art de Joliette. Léonard-Contant is a recipient of grants from the Conseil des arts et lettres du Québec and the Canada Council for the Arts. She has also been the recipient of the Antrev award, and the Kala Art Institute fellowship. She has participated in several art residencies such as the Christoph Merian foundation (a Quebec studio in Basel); Omi International residency, Ghent; Residency Unlimited, New York; and the Banff Art Centre, among others. She graduated with an MFA from Concordia University in 2012.

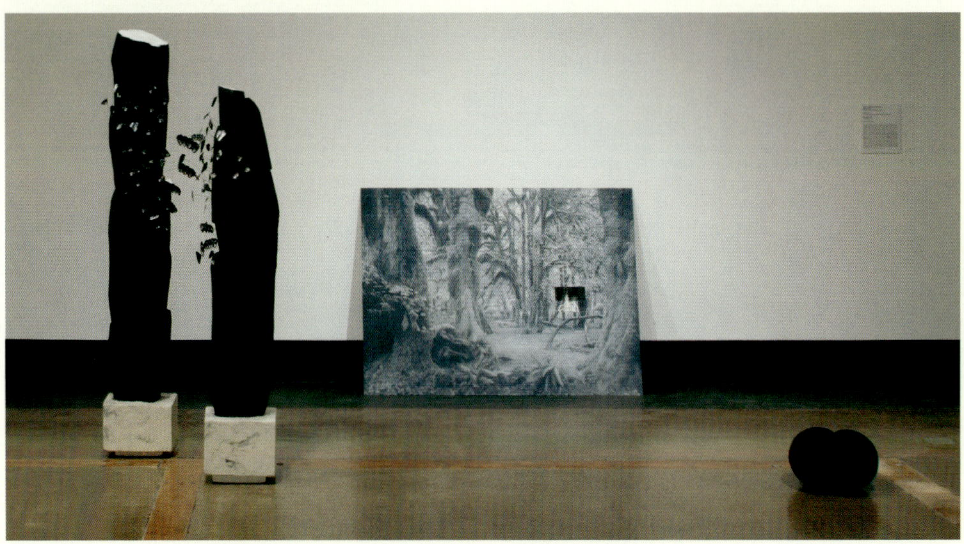

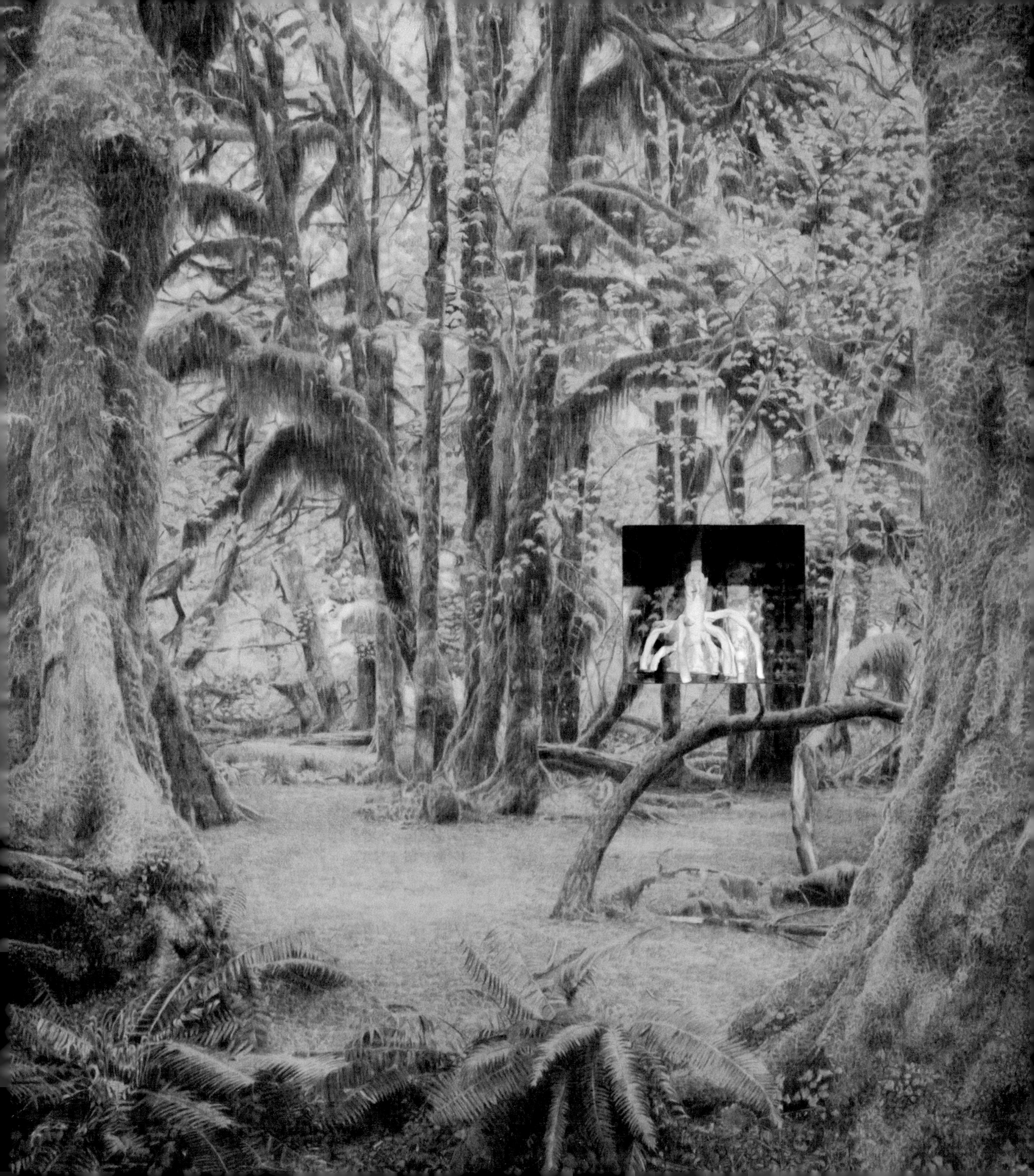

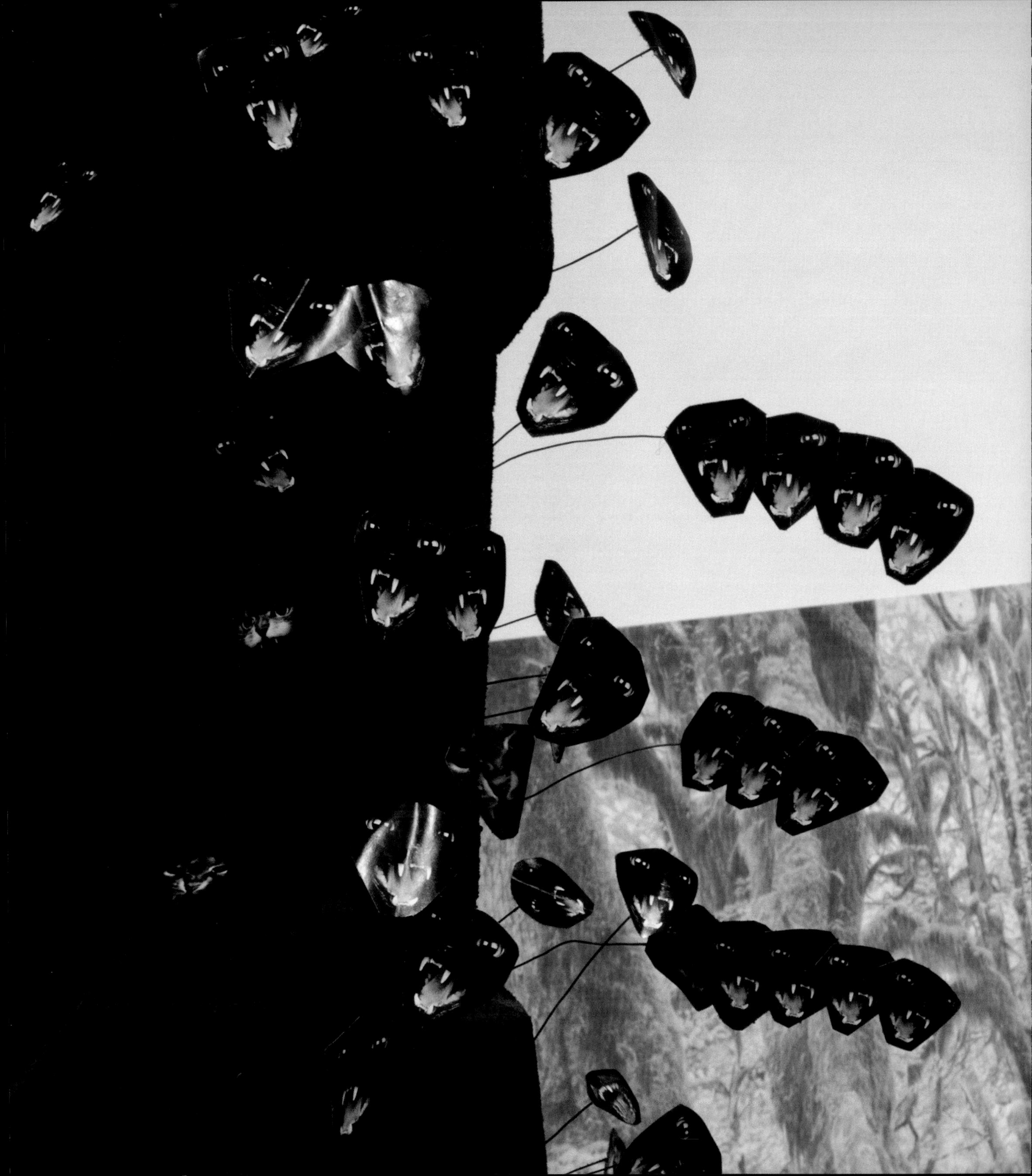

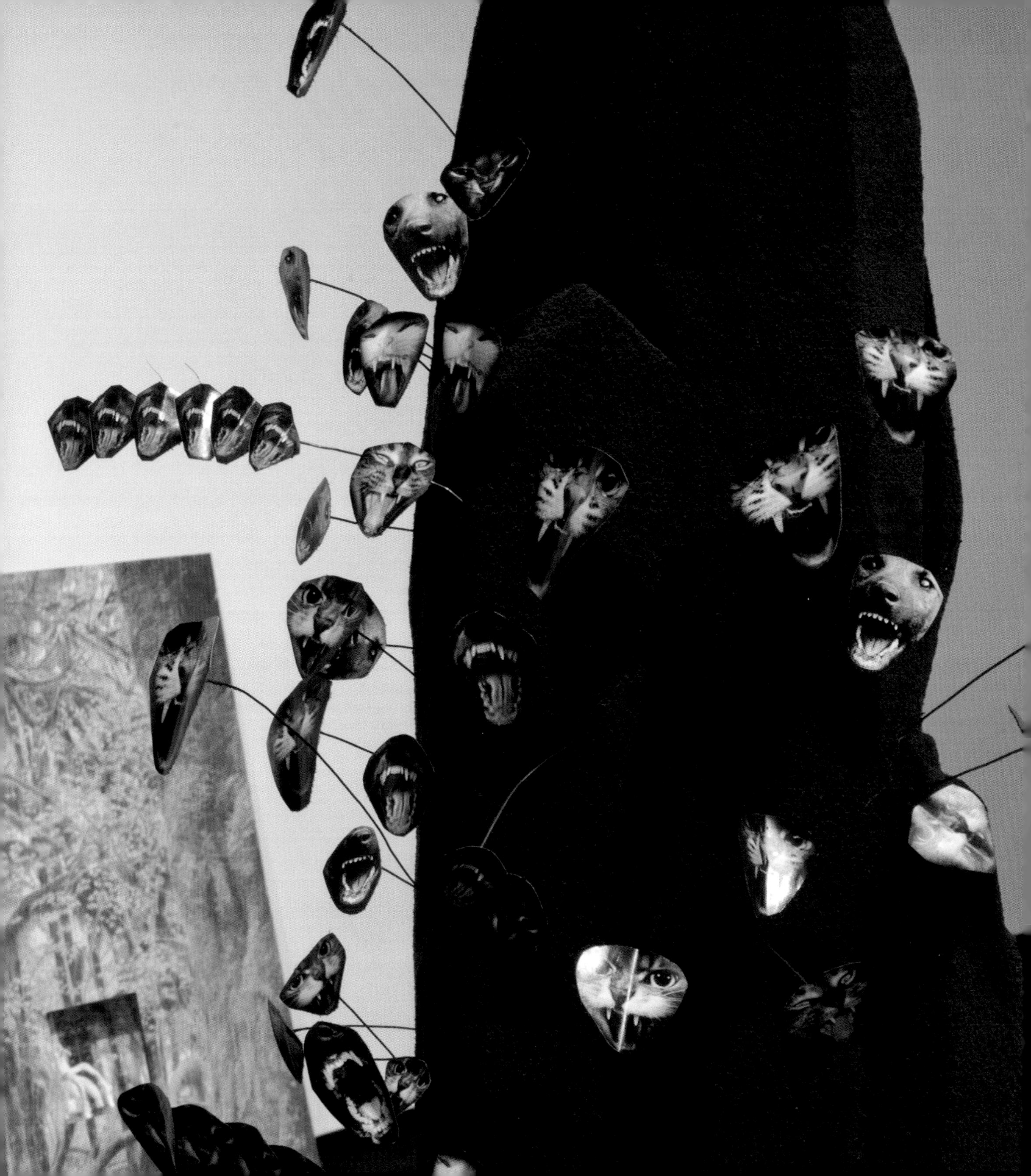

Lynne Marsh

OPPOSITE and FOLLOWING PAGES
***Plänterwald*, 2010**
High-definition video, 17:50, video projection with 4 channel sound, raised wood screen construction. Courtesy of the artist and Galerie Donald Browne, Montreal

Plänterwald is filmed on the site of a former German Democratic Republic amusement park built in 1969 and abandoned after unification. Its rollercoaster and Ferris wheel sit motionless at the edge of the city of Berlin. After being closed to the public for almost a decade the rides and fairground structures—once providing a distraction from everyday realities—have been left to a gradual process of decay and overgrowth. Paradoxically, this derelict site is patrolled and protected by security guards who on the one hand attempt to maintain its separation from the public sphere and contemporary life yet on the other hand position it in the present social and economic conditions. The video stages a journey in, over, and through this bordered-off park, evoking the exceptional conditions of its persistent existence. Positioning the security guards as the guardians of a "dead" space, the work plays on the absurdity of the use of proprietary force over an otherwise relinquished site. *Plänterwald* pursues an exploration of a world held together by an internal logic, and quietly, yet relentlessly—like the defunct rollercoaster—echoes the rumbles of deep social and political fault lines and their explosive potential.

Lynne Marsh's (b. 1969 in Vancouver, British Columbia) practice lies at the intersection of moving image, installation and performance. Her works address the latent spaces of spectacle by exploring behind-the-scenes production and the absent audience. Marsh engages specific sites and architectures that bear a conflation of historical, social and political forces and that function as stages where the performative is derived from the organizational and physical structures already at play. Her research is fuelled by a reflection on how these social spaces and their ideological orientation can be reconfigured through the camera lens. She directs the movement of bodies and cameras in parallel, thus re-inscribing the performative onto the apparatus of capture and shifting political readings of social sites. What emerges is a complex relationship between support and participation, camera and subject, and the individual and the social. Lynne Marsh received her BFA from Concordia University, Montreal in 1992 and her MFA from Goldsmiths College, London in 1998. She exhibits internationally with recent solo shows at the Musée d'art contemporain de Montréal, Danielle Arnaud contemporary art in London, and PROGRAM in Berlin, and group exhibitions that include the 10th International Istanbul Biennial; Centro Cultural Montehermoso, Spain; Manif d'art 5, Quebec City, Canada; Oakville Galleries, Canada; 53 Art Museum in Guangzhou, China; and The National Gallery of Canada. She divides her time between Montreal, Berlin and London.

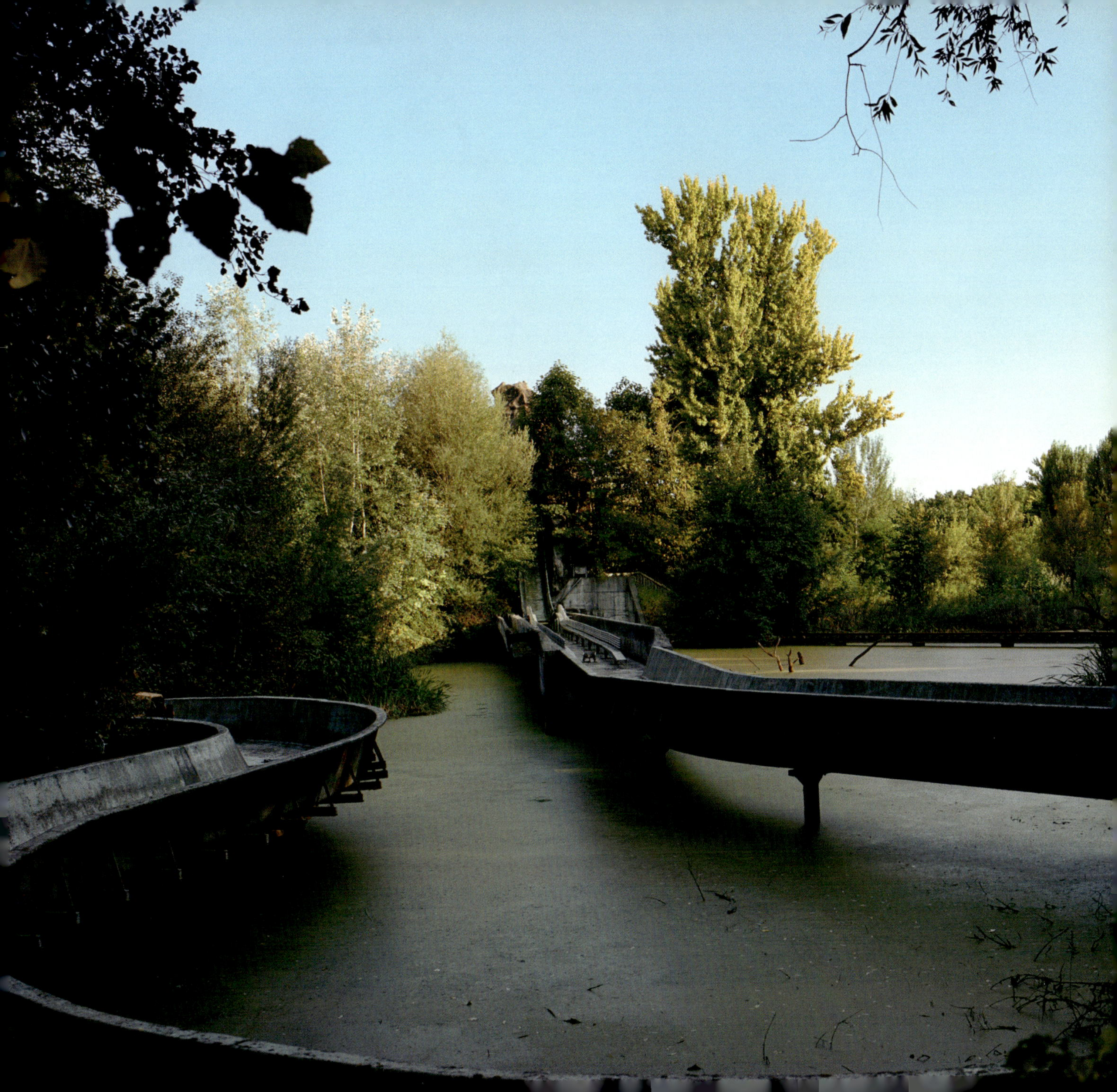

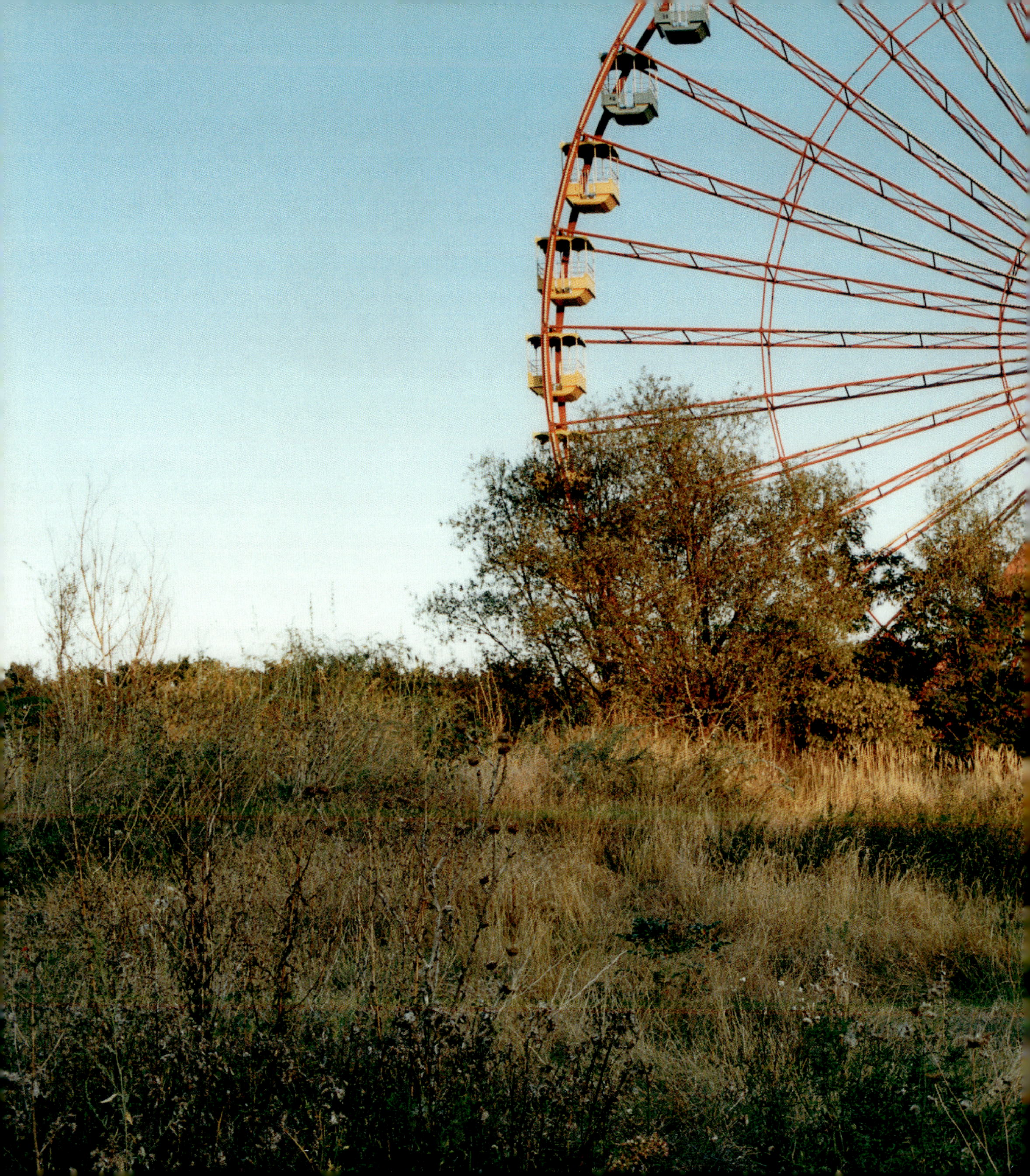

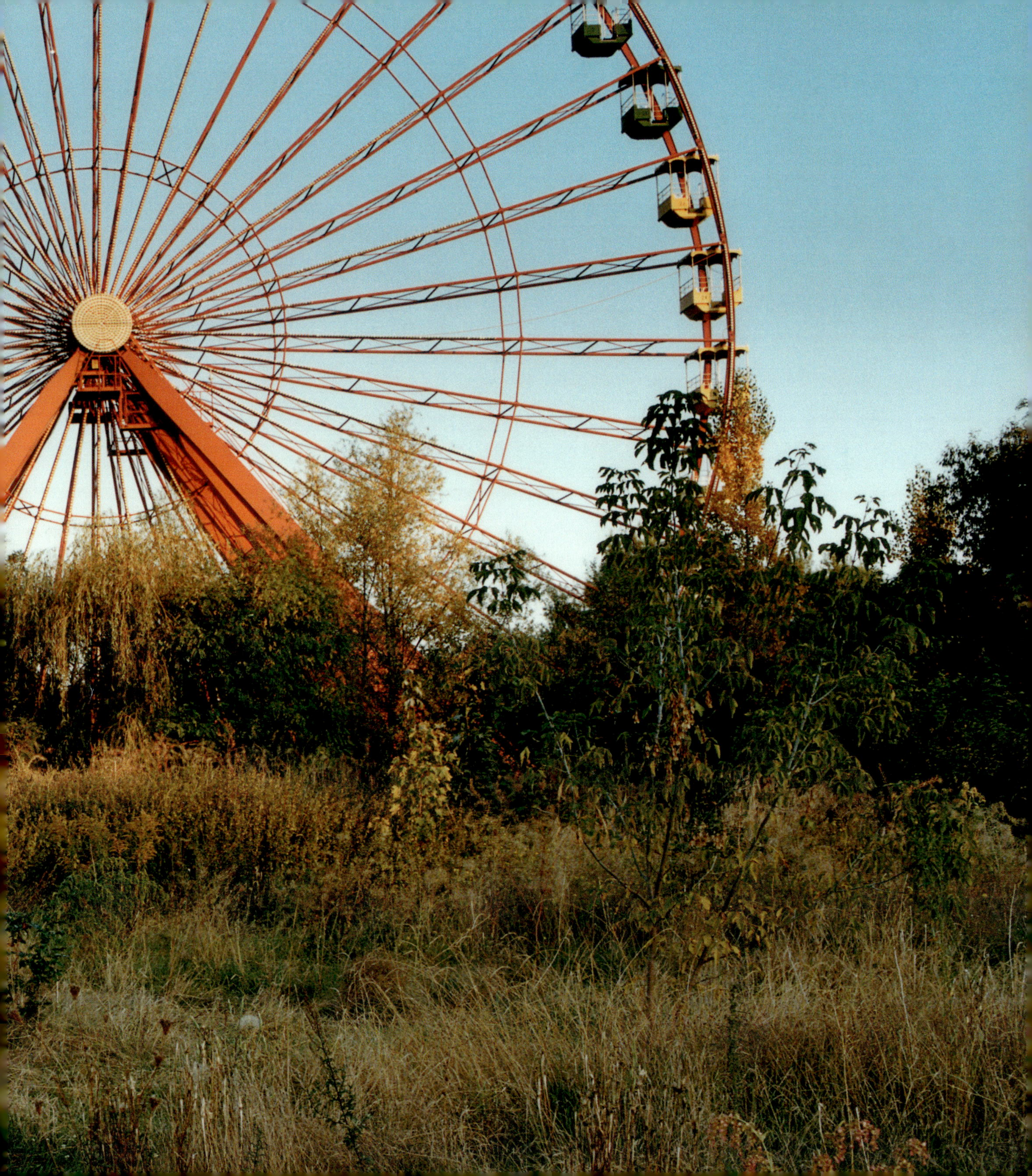

Lisa Sanditz

OPPOSITE
***Tang Factory I*, 2008**
Acrylic on canvas
30 x 40 inches

FOLLOWING PAGES, CLOCKWISE FROM LEFT
***Sock City*, 2005**
Acrylic on canvas
40 x 30 inches

***Lamp City*, 2009**
Acrylic on canvas
68 x 87 inches

***Tower of Babble*, 2009**
Acrylic on canvas
30 x 40 inches

Courtesy of the artist, CRG Gallery, New York; and Galerie Rodolphe Janssen, Brussels

1. David Barboza, "In Roaring China, Sweaters are West of Socks City," *New York Times*, 24 December 2004, http://www.nytimes.com/2004/12/24/business/worldbusiness/24china.html?_r=0 (accessed January 2013).

For the last ten years, Sanditz's paintings have focused on the various forms in which the marketplace and wilderness intersect, overlap and inform each other. *Sock City* was the beginning of a series of paintings by Sanditz based on industrial cities in China. The artist had read an article in the *New York Times* in 2004 about Sock City, and other single industry cities in China—cities that were nicknamed after the fantastic volume of production undertaken there.[1] Sock City was one of the first places she visited the first time she went to China.

Lamp City is a painting of the monumental hurricane lamp proposed for the city square of Lamp City, Zhongshan, China. The suggested monument is a towering glass skyscraper, but due to lack of funds, the actual monument is a low-rise concrete sculpture. After visiting the Shanghai Museum for the first time, Sanditz considered the institution's architecture—which emulates the shape of a Tang Dynasty vessel—and the extensive market for reproductions of Chinese pottery. She made this painting based on the industrial area outside of Shanghai where much of this pottery is produced. She limited the palette of the painting to the gold, green and brown palette of the Tang pottery and used polyurethane to create a glaze-like effect in the painting, akin to that of the pottery.

Tower of Babble was painted after the artist took a trip to the world's largest commodities market in Yiwu, China. The Tower is a wholesale market the size of the Empire State Building on its side (43,000,000 square feet) filled with every conceivable thing a person could buy. The title *Tower of Babble* is drawn from a billboard advertising a dazzling variety of cellphones near the market. Sanditz is examining how the consumer-based relationship between China and the United States is causing the Chinese landscape to shift from the pastoral to the industrial.

Lisa Sanditz (b. 1973 in St. Louis, Missouri) works in the Hudson Valley in New York. Her painting subjects have included sporting events, shopping malls, residential developments and tourist destinations. Through a complexity of combined forms and the clash of unconventional and more historically-anchored painting techniques, Sanditz has described the brash and metastasizing flux of the American landscape. Her continued investigation of these places has yielded an awareness of multinational corporations and sprawling urban and exurban areas not just in her native America, but all over the world. She will have her fourth solo show at CRG Gallery, New York, in 2014, and she has recently had work in the Weatherspoon Drawing Biennial in Greensboro, North Carolina. She has had solo exhibitions at ACME Gallery, Los Angeles; Galerie Rodolphe Jansen, Brussels; and The Kemper Museum of Contemporary Art in Kansas City. She has participated in group exhibitions at The Torrance Art Museum in California; The Nerman Museum of Contemporary Art in Overland Park, Kansas; Halsey McKay Gallery in East Hampton, New York; The American Academy in Rome; The Wexner Center for the Arts in Columbus, Ohio; and Contemporary Art Museum Houston (CAMH), Texas. Her work has been reviewed in the *New York Times*, the *New Yorker*, and *Art Forum* among other publications. She recently worked on a commission for the Johns Hopkins Children's Center in Baltimore, Maryland. She was a Guggenheim Fellow recipient in 2008, which helped support the work produced for *Ecotopia*. She is an Assistant Professor at Bard College, New York.

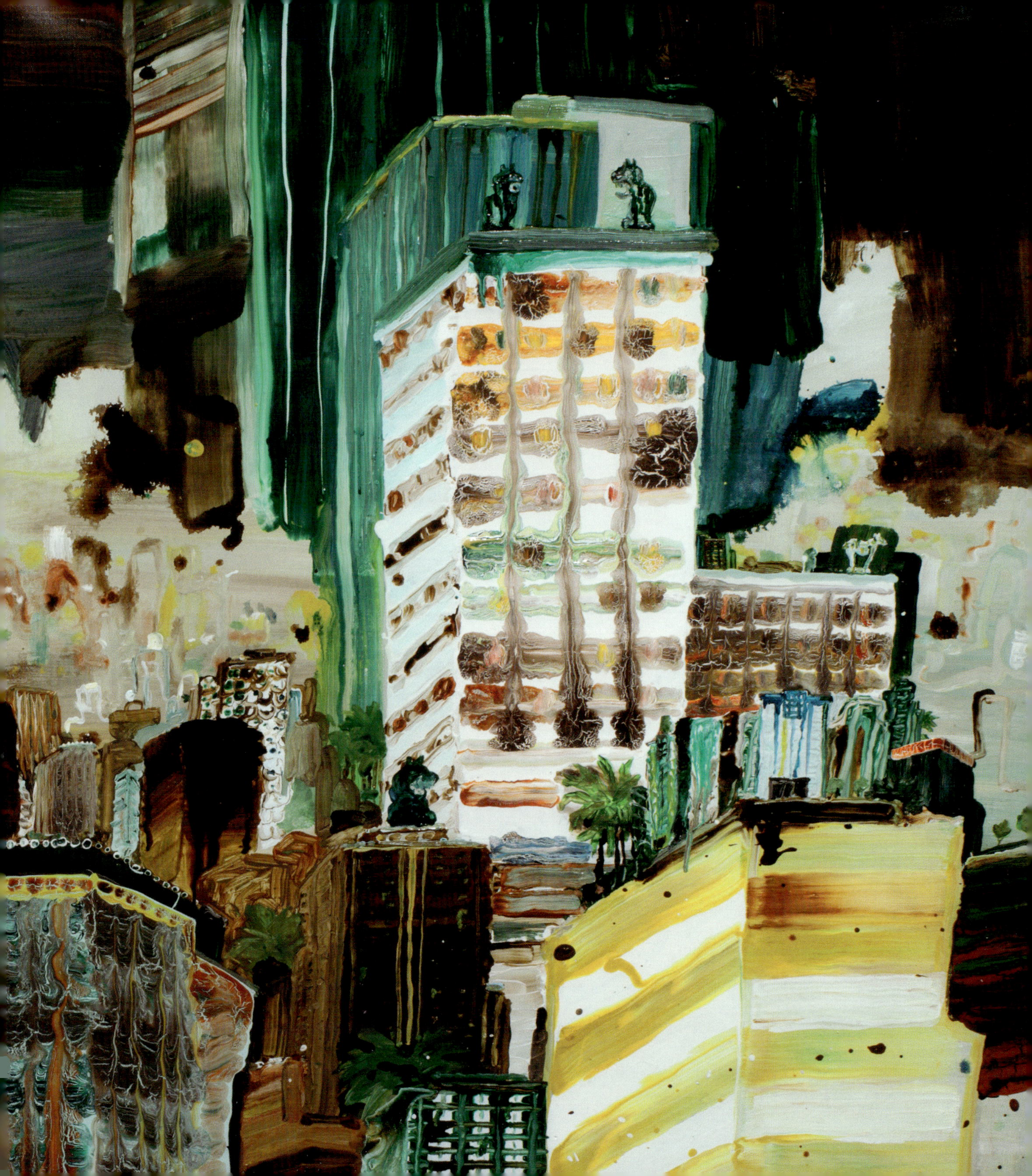

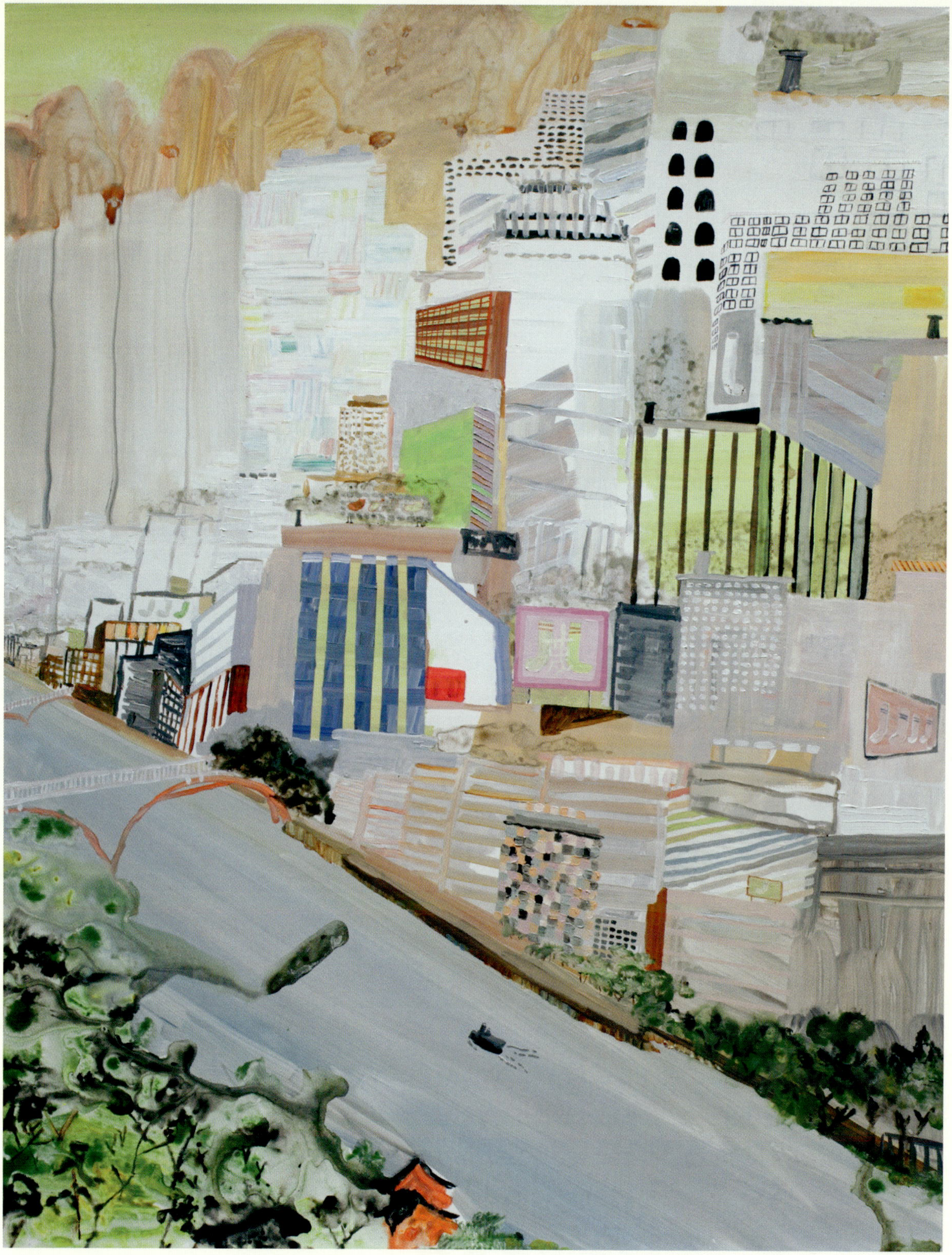

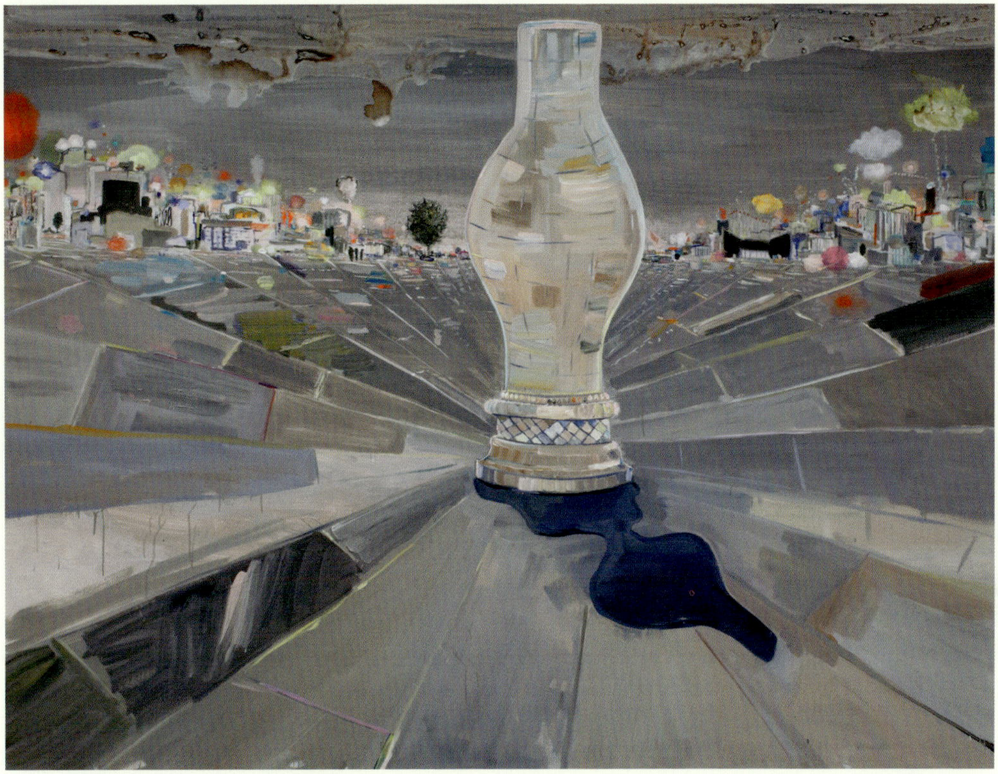

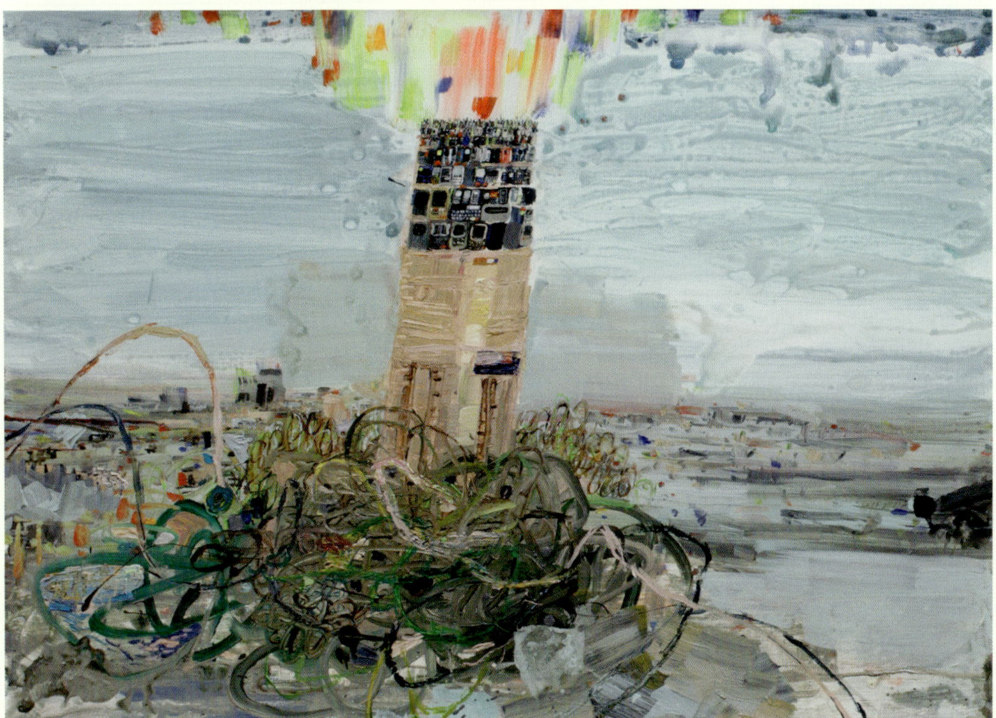

Jennifer Steinkamp

OPPOSITE and FOLLOWING PAGES
***Orbit 6*, 2010**
Video installation
Dimensions variable, edition of 3, approximately 4 minutes. Courtesy of the artist and Lehmann Maupin Gallery, New York

1. Text supplied by Lehmann Maupin Gallery, New York, August, 2012
2. Didactic text from a Jennifer Steinkamp exhibition at The Museum of Fine Arts, Houston (2012), http://www.mfah.org/exhibitions/past/jennifer-steinkamp/ (accessed August 2012).

Orbit 6 is a digital animation depicting a year through the changing of four seasons in an endless four minute loop. The shift in leaf colours and the motion of various fictional trees across a black background indicate the seasonal transitions. The title of the work references an orbital path, the motion of a planet around a star that determines a year.[1] Steinkamp builds her work around the way that light can dematerialize space, indoors and out. She uses 3-D digital animation software to create video projections that are at once vividly theatrical and seductively illusionistic. Steinkamp fabricates fantastically beautiful animations that could not exist in the real world, charting the cycles of the seasons and the passage of time.[2]

Jennifer Steinkamp (b. 1958 in Denver, Colorado) is a Los Angeles-based artist who employs computer animation and new media to create projection installations in order to explore ideas about architectural space, motion, and phenomenological perception. Her digitally animated works make use of the interplay between actual space and illusionistic space, thus creating environments in which the roles of the viewing subjects and the art objects become blurred. Recently, Steinkamp has had solo exhibitions of her work at Greengrassi, London; Fabric Workshop and Museum, Philadelphia; ACME, Los Angeles; VCUarts Anderson Gallery, Richmond, Virginia; and Galería Soledad Lorenzo, Madrid. Other recent projects and exhibitions include *Five in Istanbul: A Selection of Artists from Lehmann Maupin Gallery* at the Borusan Müzik Evi in Istanbul, Turkey; *Jennifer Steinkamp: Madame Curie* at the Museum of Contemporary Art San Diego; set design for Arnold Schoenberg's *Erwartung* at the New York City Opera; and *Prospect. 2 New Orleans* at the New Orleans Museum of Art. Her work can be seen in numerous public collections, including the Los Angeles County Museum of Art, the Corcoran Gallery of Art in Washington DC, the Hammer Museum in Los Angeles, the Albright-Knox Gallery in Buffalo, Istanbul Modern in Turkey, the Towada Art Center in Japan, and in important private collections worldwide.

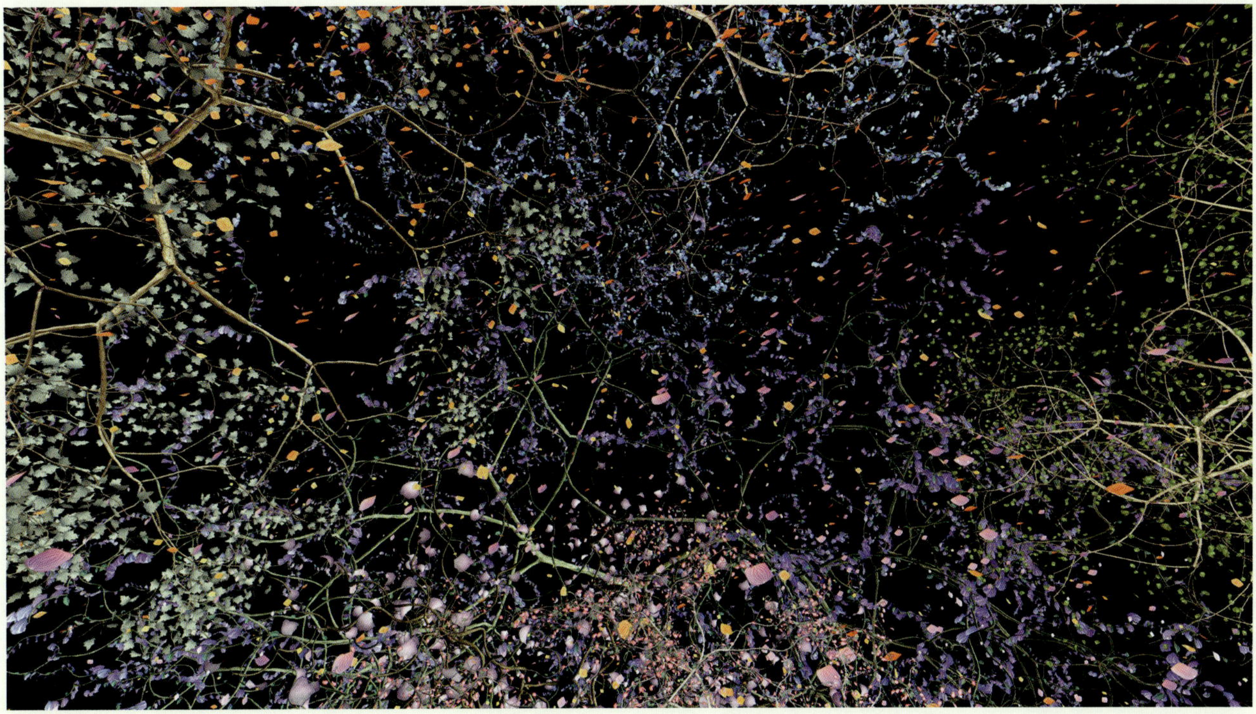

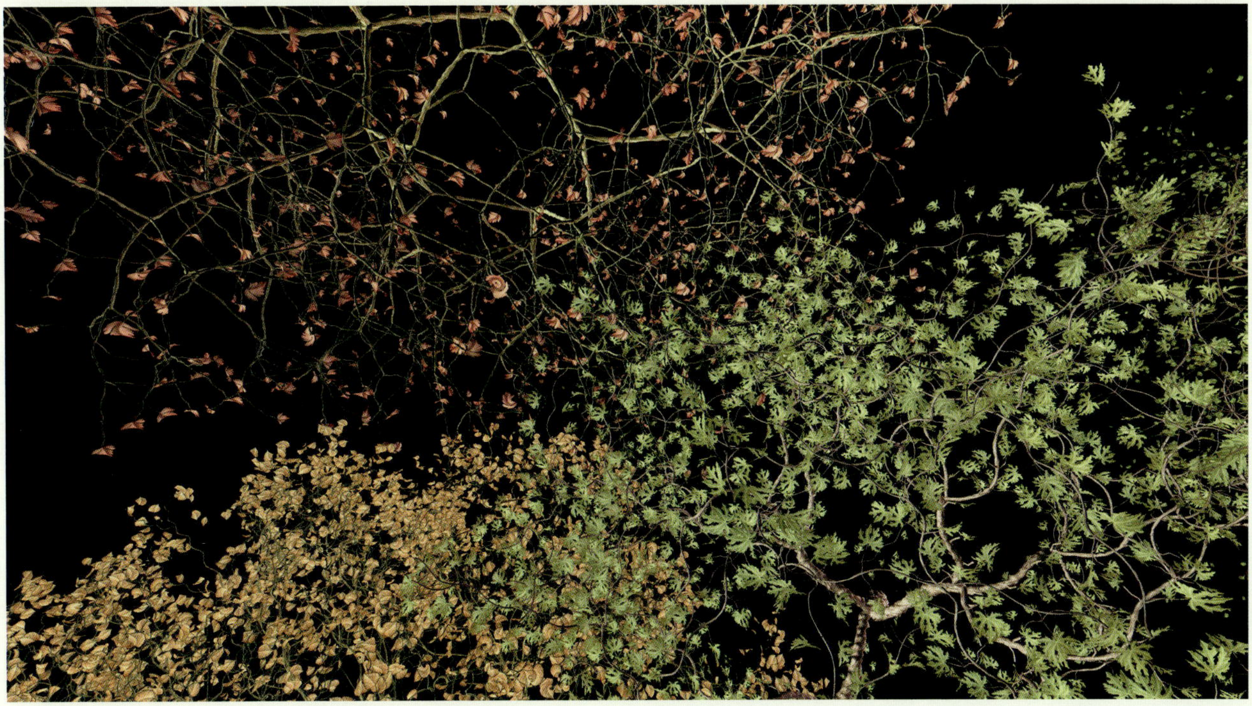

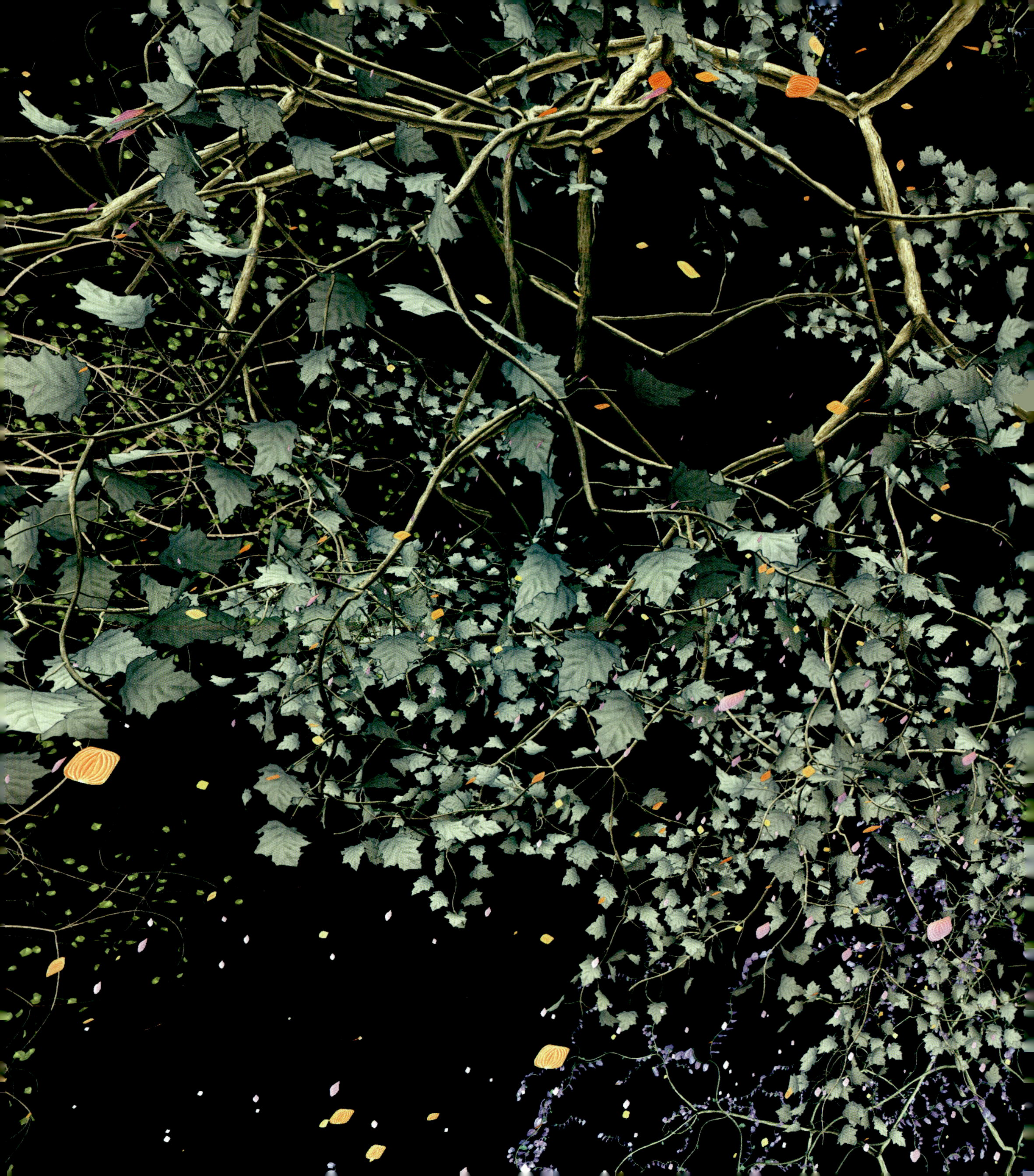

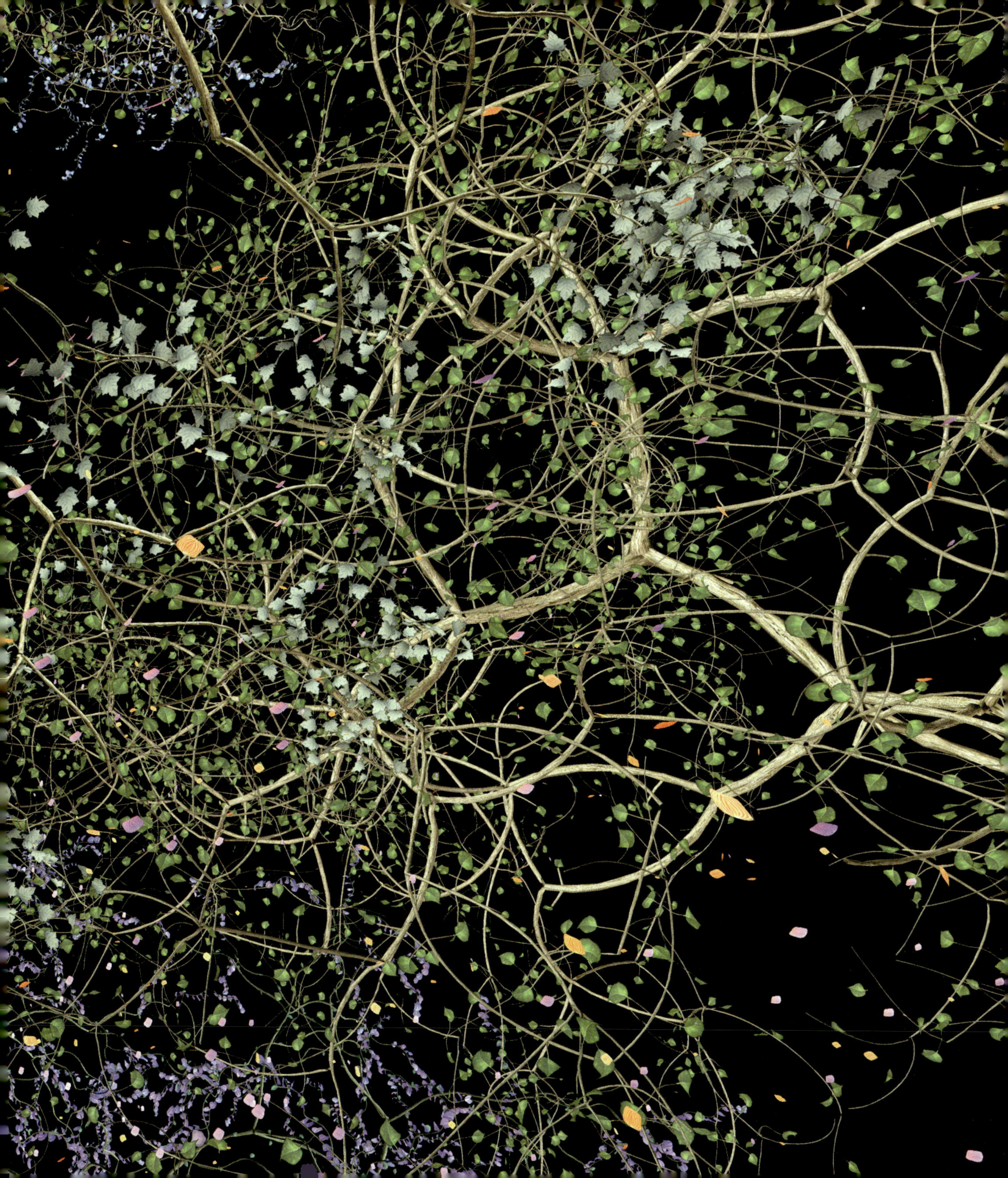

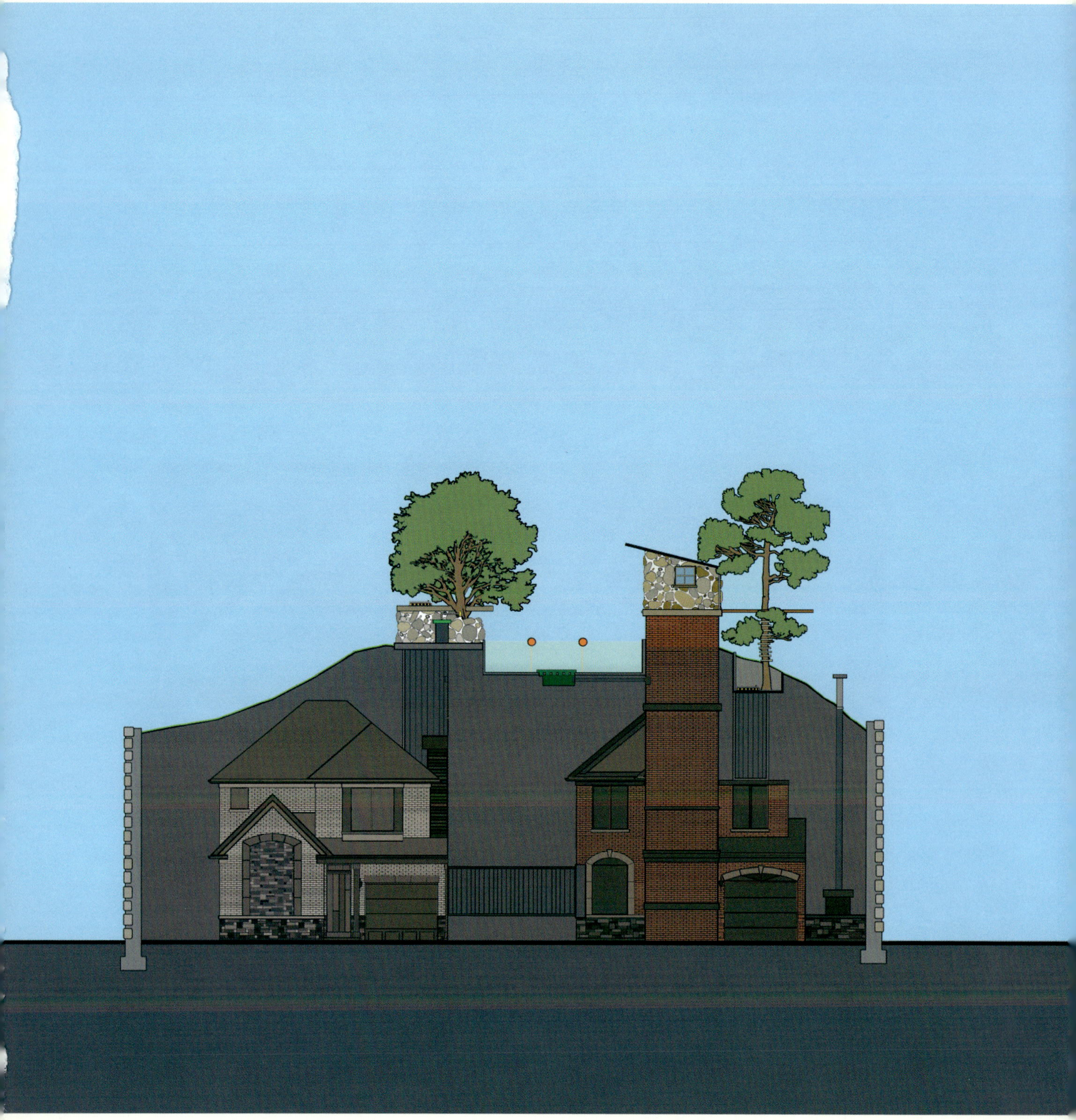

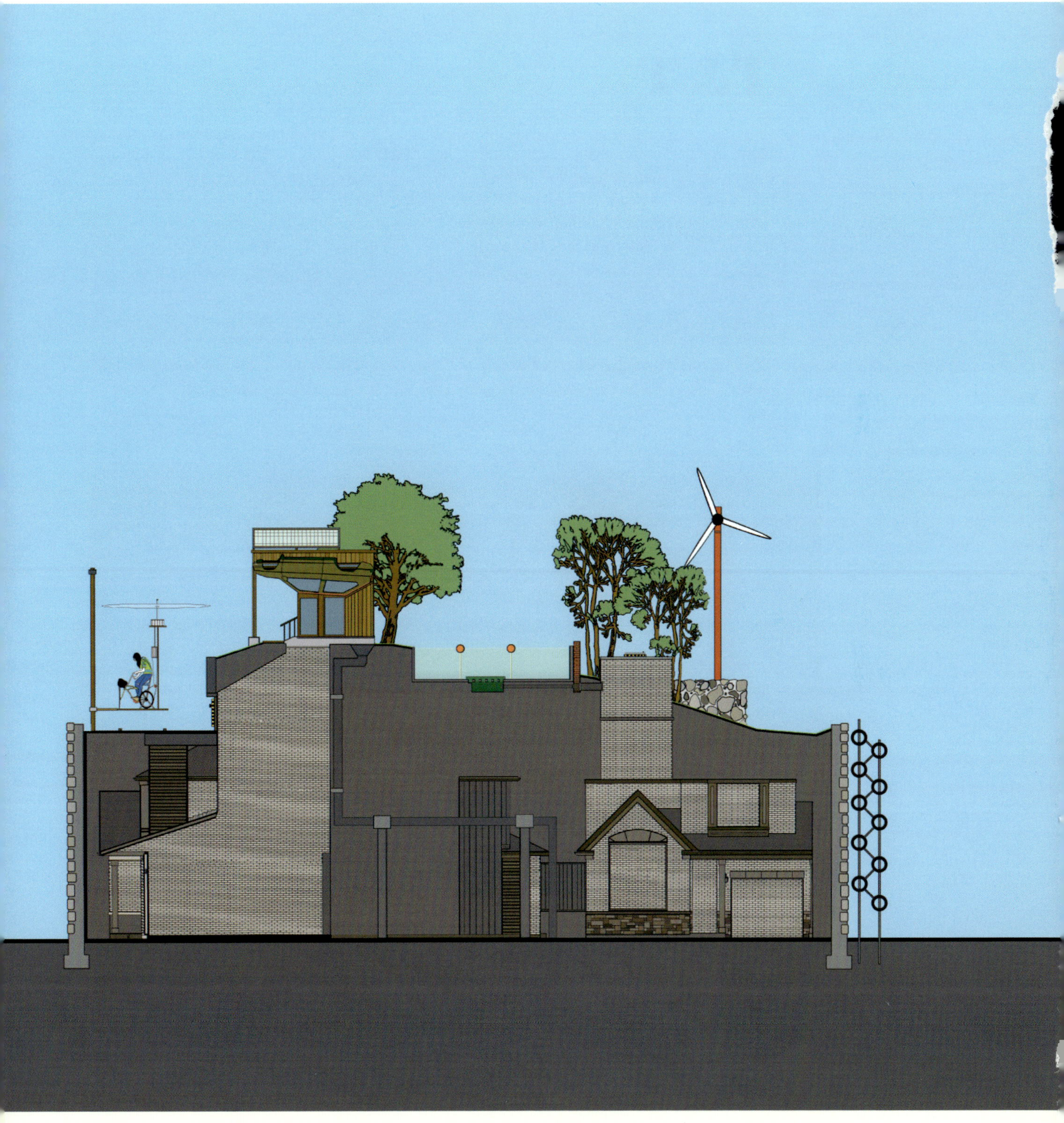

T&T

OPPOSITE and
FOLLOWING PAGES
***Wilmont Trail*, 2003**
Digital C-print, 94 x 24 inches. Courtesy of the artists and Trapp Editions, Vancouver

1. Patrik Andersson, "Doubt/Hope" in *Onward Future*, (Toronto and London: Oakville Galleries and Museum London, Canada, 2008), 10.
2. Betty Ann Jordan, "Tony Romano zand Tyler Brett: Carchitecture, and Cautionary Ingenuity," *Canadian Art* online, 18 December 2008, http://www.canadianart.ca/reviews/2008/12/18/romano-brett/ (accessed August 2012).

In the *Onward Future* exhibition catalogue, curator Patrik Andersson said, "Suspending us in the present between the past and the future, childhood and adulthood, reality and fiction, T&T (Tyler Brett and Tony Romano) locates a heterotopia sustained by imagination, innovation and play."[1] Writer Betty Ann Jordan describes their collaborative work as "Photoshopped renderings...that imagine off-the-grid survival after an apocalyptic disaster...Sparse and spindly coniferous trees hint at massive deforestation while doubling as living structural uprights. Small crews of resourceful back-to-the-landers generate heat and light by harnessing hydro, solar, wind and geothermal power."[2] Humans work in synthesis with nature in T&T's images, creating new means of ecologically sustainable living. *Wilmont Trail* depicts a unique form of architecture that combines bricks, wood and trees, and depicts humans standing atop tall platforms, surveying the brave new world before them while acquiring new skills and methods for survival. It captures the hopeful possibilities for how our attitudes toward the environment can be re-envisioned.

T&T (Tyler Brett, b. 1970 in Chilliwack, British Columbia and Tony Romano, b. 1978 in Toronto, Ontario) has been working collaboratively on images, models, sculptures, and sound works since 2001. Many of its concepts are based on re-engineering, retrofitting, and combining everyday objects and architecture with sound ecological principles, imaginary technologies, and social harmony. Tony Romano graduated with a BFA from the Emily Carr Institute of Art & Design, Vancouver in 2001. He is based in Toronto, Ontario where he and artist Claire Greenshaw co-edit and design the Canadian arts and culture magazine, *Millions*. Tyler Brett graduated with a BFA from the Emily Carr Institute of Art & Design, Vancouver in 2001 and completed his MFA at the University of Saskatchewan in 2009. Brett is based in Bruno, Saskatchewan where he and artist Kerri Reid co-direct the Bruno Arts Bank artist residency, gallery, mini-museum and presentation venue. T&T has exhibited work both nationally and internationally and is represented by Clint Roenisch Gallery in Toronto and Trapp Editions in Vancouver.

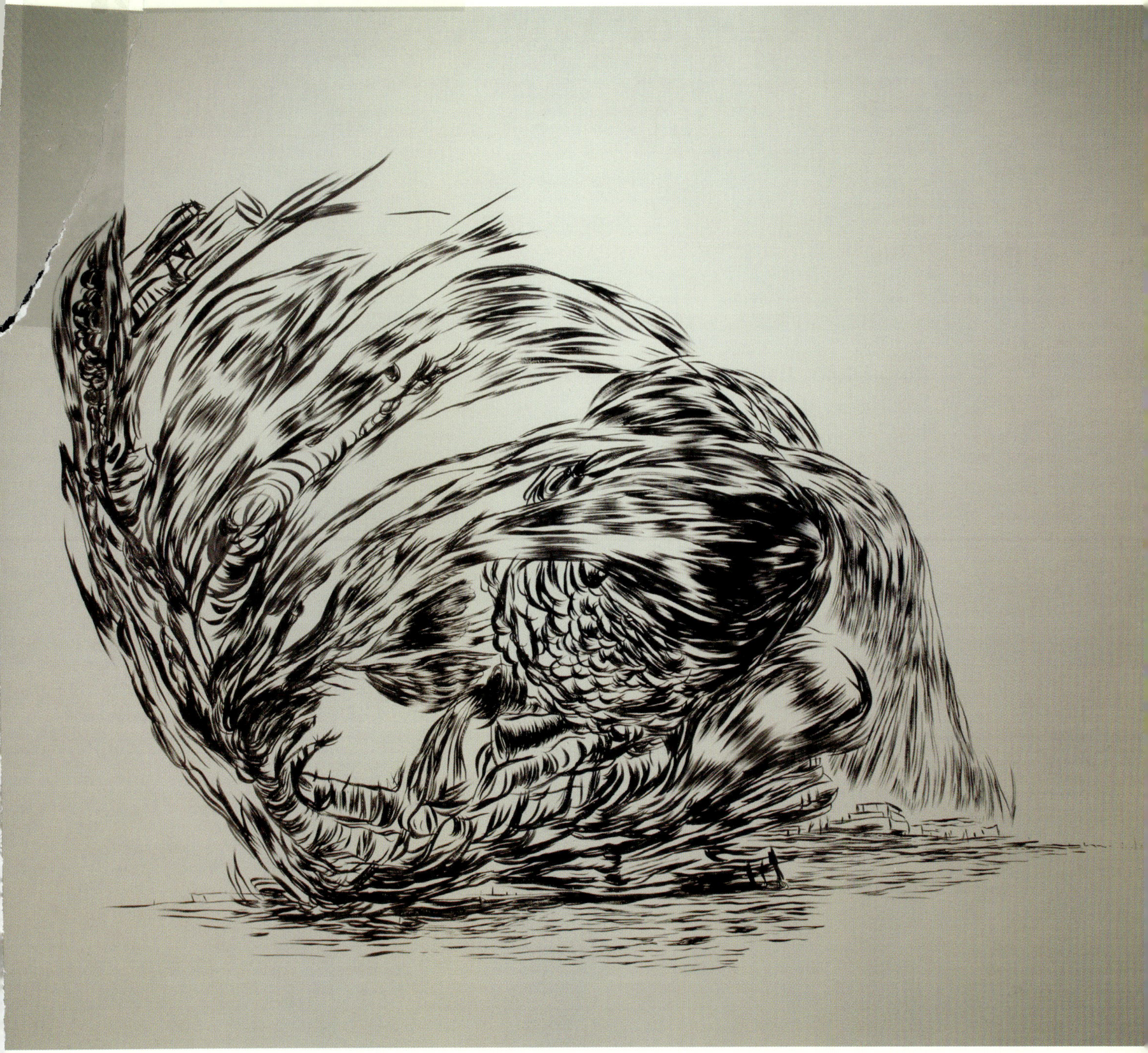

List of Works

BGL	*Pinocchio*, 2009 Chainsaw, clamp, plywood, vinyl, paint; with pair of speaker fans (fans, wood, plastic) 112 x 115 x 126 inches; each fan: 16 x 32 x 19 inches Courtesy of the artists and Diaz Contemporary, Toronto		
David Brooks	*Unfinished Section of Sidewalk with Two Palms and Rebar*, 2012 Concrete, rebar, plant material 30 x 22 x 36 inches Courtesy of the artist and American Contemporary, New York		
Dagmara Genda	*Panorama*, 2012 Collage with ink, latex, and acrylic on paper 52.5 x 301.4 inches Courtesy of the artist		
Rodney Graham	*Ponderosa Pines, Princeton B.C./ CAT hi-way yellow*, 1992 Edition 5/15, 5 chromogenic prints Each 10 x 8 inches Collection of McCarthy Tétrault, LLP, Toronto		
Isabelle Hayeur	*Ascendance*, 2005–2008 Inkjet on polyester 125 x 39 inches	*Losing Ground*, 2009 High-definition video 13:00	
	Courtesy of the artist and Galerie Division, Montreal.		
Tristram Lansdowne	*Envelope*, 2012 Watercolour on paper 36 x 25.5 inches Courtesy of the artist	*The Nurse*, 2011 Watercolour on paper 40 x 30.5 inches Collection of Mark Zadorozny, Toronto	
Maude Léonard-Contant	*Creeper*, 2012 Coloured crayon, mdf, plexiglas, polymer 49.6 x 38.2 x 3.9 inches	*Cacti*, 2012 Styrofoam, playdough, digital prints, flocking, wire Dimensions variable	*Coco-fesse*, 2011 Rubber foam, ink 7.9 x 11.8 x 12.6 inches
	Courtesy of the artist		

Kate Wilson

OPPOSITE
***Untitled*, 2012–2013**
From an original drawing measuring 10.6 x 13.7 inches, three sections
Ink on semi-matte paper
86.4 x 144 inches
Courtesy of the artist and General Hardware Contemporary, Toronto

Wilson's work shows a darker world littered with industrial waste, abandoned structures, and hydro poles in conjunction with a focus on the powerful forces of natural phenomena. Her miniature architectures are deeply imbedded in organic environments that themselves are caught up in wild weather. Her work also reflects her concern that while technologies such as genetic modification, monocultures, and oil extraction are devastating natural environments, natural phenomena, such as hurricanes, floods, and climate change are also radically altering our environment. In *Untitled*, Wilson has depicted a frightening scene where a botanical swirl is caught up in a hurricane-like wind, sucking up nearby architecture and other structures and elements in nature that line the horizon in this large-scale drawing. One is reminded of major catastrophic natural disasters such as Hurricane Katrina in New Orleans in 2005. What are the dystopian and utopian elements of what is left behind in an environment after a hurricane has swept through it?

Kate Wilson (b. 1955 in Toronto, Ontario) is a Toronto-based artist whose practice includes large-scale wall drawing installation, and experimental animation. She attended the Ontario College of Art & Design from 1975–1977, among other art schools and residencies. Wilson has exhibited nationally and internationally and is a recipient of awards from the Toronto Arts Council, Ontario Arts Council, and the Canada Council for the Arts. Her recent solo exhibitions include *Botanical Model City*, Museum London, Ontario; and *The Afterlife of Buildings*, General Hardware Contemporary, Toronto. Recent group exhibitions include: *Beyond/In Western New York 2007* organized by the Albright-Knox Art Gallery, Buffalo; *Canadian Club*, Canadian Cultural Centre, Paris; *Cultural, Temporal and Imagined: Landscape in Recent Contemporary Acquisitions*, Art Gallery of Greater Victoria, British Columbia; and *Toronto | Berlin*, Zweigstelle Berlin. Kate Wilson's animated film *A Primer of Small Stars* premiered at the National Film Board of Canada Mediatheque, Toronto in June 2012.

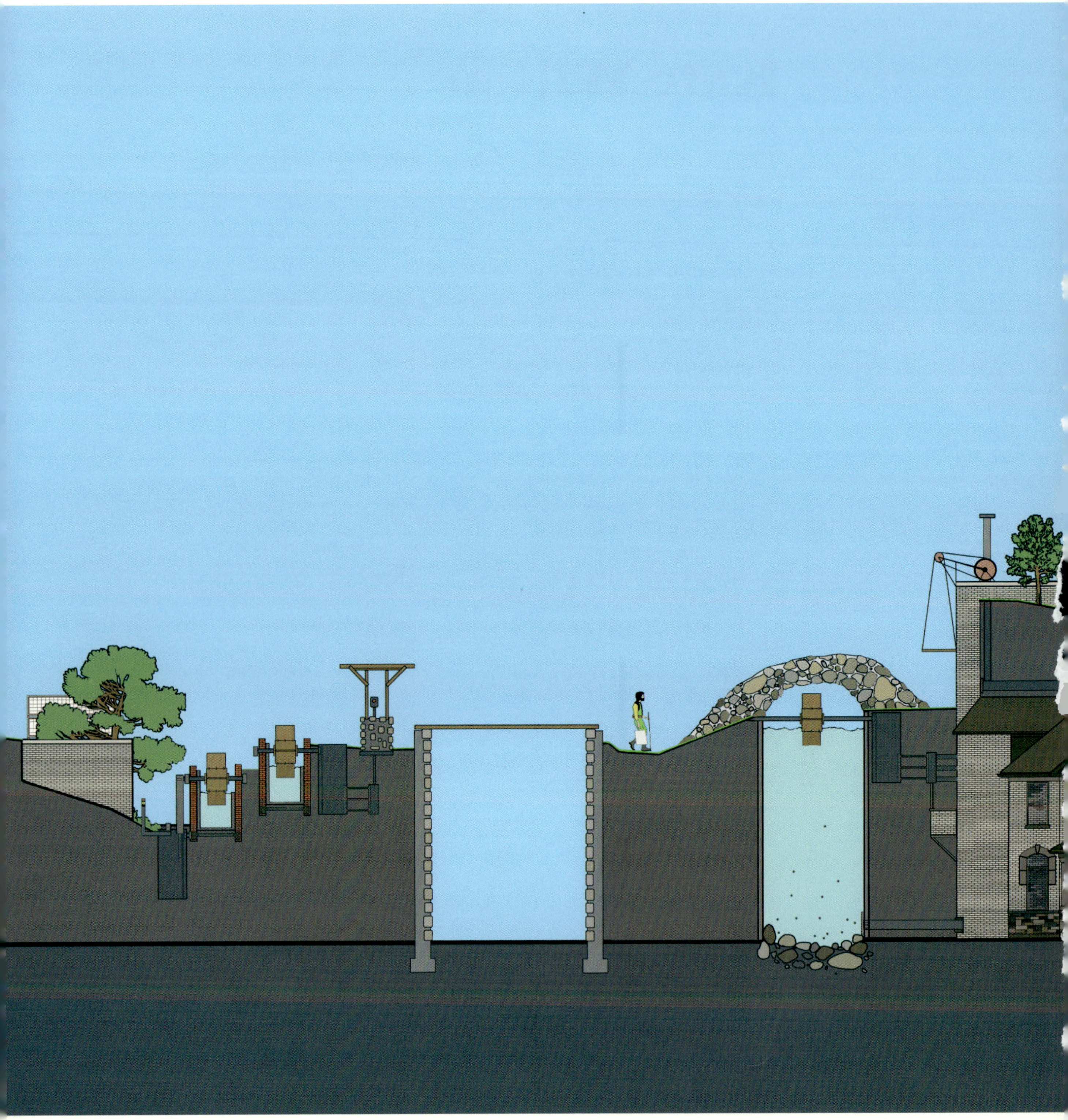

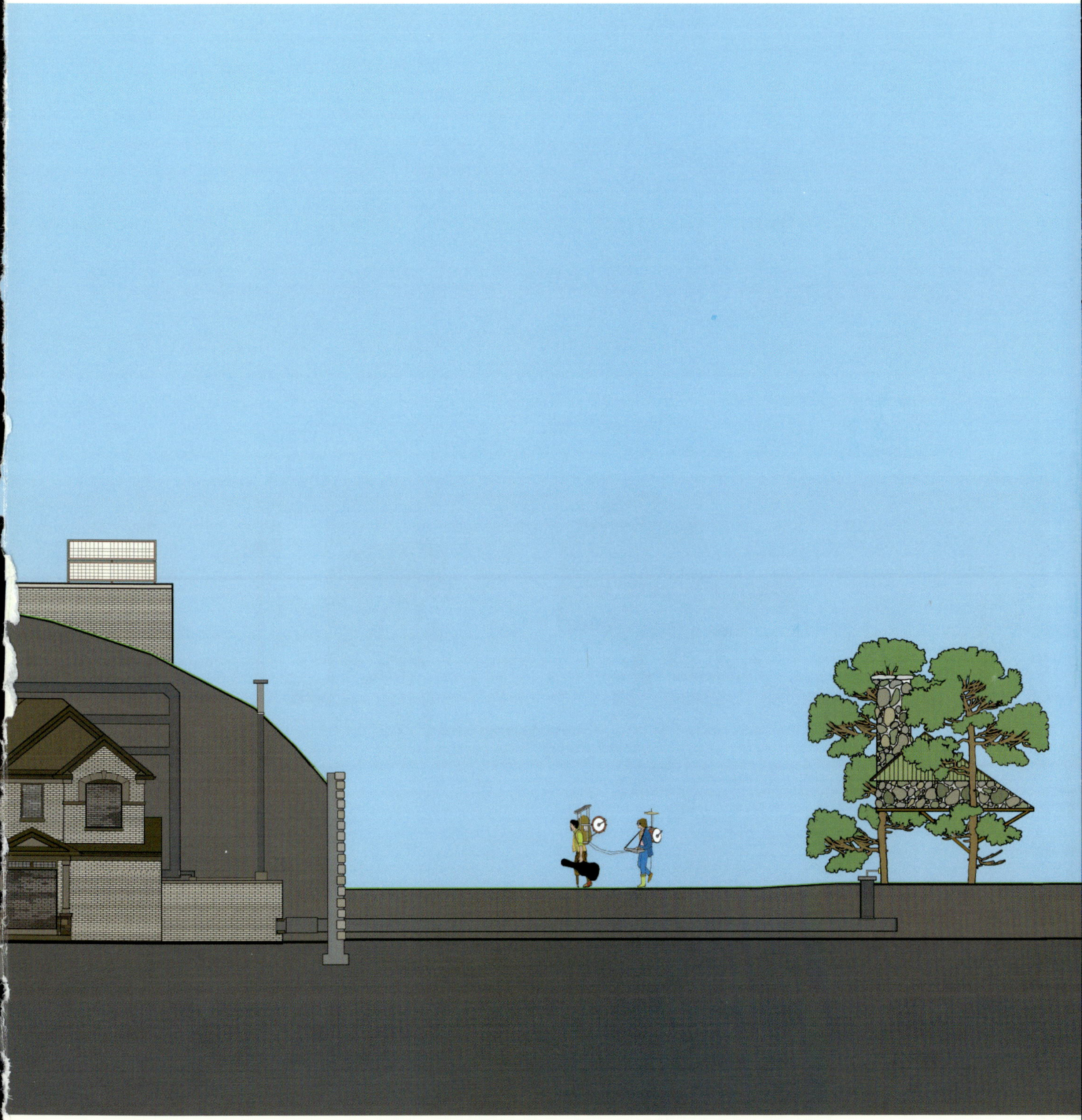

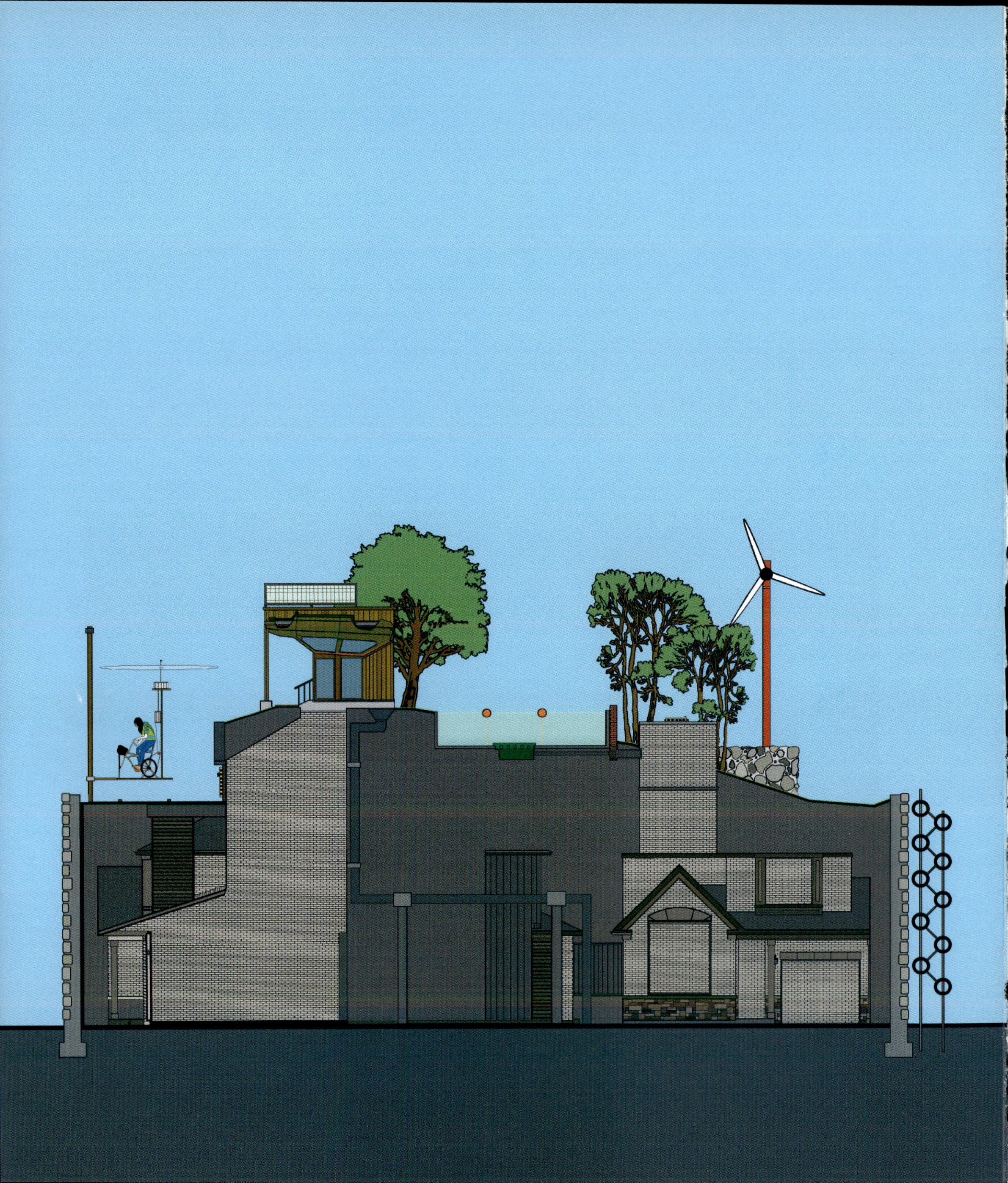

Lynne Marsh

Plänterwald, 2010
High-definition video, 17:50
Video projection with 4 channel sound, raised wood screen construction
Courtesy of the artist and Galerie Donald Browne, Montreal

Lisa Sanditz

Lamp City, 2009	*Sock City*, 2005	*Tang Factory I*, 2008	*Tower of Babble*, 2009
Acrylic on canvas	Acrylic on canvas	Acrylic on canvas	Acrylic on canvas
68 x 87 inches	40 x 30 inches	30 x 40 inches	30 x 40 inches

Courtesy of the artist, CRG Gallery, New York; and Galerie Rodolphe Janssen, Brussels

Jennifer Steinkamp

Orbit 6, 2010
Video installation
Approximately 4 minutes
Dimensions variable, edition of 3,
Courtesy of the artist and Lehmann Maupin Gallery, New York

T&T

Wilmont Trail, 2003
Digital C-print
94 x 24 inches
Courtesy of the artists and Trapp Editions, Vancouver

Kate Wilson

Untitled, 2012–2013
From an original drawing measuring 10.6 x 13.7 inches, three sections
Ink on semi-matte paper
86.4 x 144 inches
Courtesy of the artist and General Hardware Contemporary, Toronto

Acknowledgements

I owe my gratitude first and foremost to the dedicated staff of the Kitchener-Waterloo Art Gallery, namely Shirley Madill, Executive Director; Crystal Mowry, Senior Curator; Barbara Hobot, former Curatorial Assistant; Linda Perez, Curatorial Assistant; Nicole Neufeld, Director of Public Programs; Jennifer Bullock, Assistant Curator and Registrar, and Teresa Chiavaroli, Communications Coordinator. I also extend my thanks to all of the talented installation team, comprised of Robert Achtemichuk, Jeff Christie, Patrick Cull, Marta Orlowska, Joshua Peressotti, and Preparator Ian Newton. I am also extremely grateful to the exhibition funders, including the Canada Council for the Arts, the Ontario Arts Council and The Kitchener and Waterloo Community Foundation. Robert McNair provided photographic documentation of the exhibition and Matthew Dupuis provided visionary graphic design of the catalogue. I also extend gratitude to videographer Scott McGovern and editor Jenn E. Norton for the introductory video that accompanies the exhibition. I would also like to acknowledge the hard work of Shannon Anderson as copy editor of this publication – her work was timely, professional, generous and critical.

Thank you to the lenders and galleries who have assisted in realizing this exhibition, including Diaz Contemporary, Toronto; American Contemporary, New York; McCarthy Tétrault, LLP, Toronto; Galerie Division, Montreal; Mark Zadorozny, Toronto; Galerie Donald Browne, Montreal; CRG Gallery, New York; Galerie Rodolphe Janssen, Brussels; Lehmann Maupin Gallery, New York; Trapp Editions, Vancouver; 303 Gallery, New York; and General Hardware Contemporary, Toronto.

This exhibition would not have been possible without the commitment and dedication of each of the artists, all of whom I met at some point during the last two years as the exhibition evolved: BGL, David Brooks, Dagmara Genda, Rodney Graham, Isabelle Hayeur, Tristram Lansdowne, Maude Léonard-Contant, Lynne Marsh, Lisa Sanditz, Jennifer Steinkamp, T&T and Kate Wilson.

I also wish to thank the Dunlop Art Gallery and the Regina Public Library, in Regina, Saskatchewan for providing the initial support in my vision for this exhibition during my tenure as Director/Curator and the hosting institutions, the Southern Alberta Art Gallery, Nickle Galleries, and the Kenderdine Art Gallery | College Art Galleries at the University of Saskatchewan, who have provided an extended life for this project.

Finally, my partner Ryan Gambrell continues to support and strengthen all the work that I do. I am so grateful for his love and encouragement.

Amanda Cachia
Exhibition Curator

KITCHENER-WATERLOO ART GALLERY

SAAG
SOUTHERN ALBERTA ART GALLERY

NICKLE GALLERIES
UNIVERSITY OF CALGARY

KAG CAG
KENDERDINE ART GALLERY
COLLEGE ART GALLERIES

Ecotopia Exhibition Sponsor:
The Kitchener and Waterloo Community Foundation -The Musagetes Fund

Canada Council for the Arts
Conseil des Arts du Canada

ONTARIO ARTS COUNCIL
CONSEIL DES ARTS DE L'ONTARIO

City of Lethbridge

Alberta Foundation for the Arts

SINCE 1948
SASKATCHEWAN ARTS BOARD

SaskCulture